Of a Different Mind

Of a Different Mind

An Introduction to the Philosophy of Mind and Psychology

Erik Myin

The MIT Press
Cambridge, Massachusetts
London, England

The MIT Press
Massachusetts Institute of Technology
77 Massachusetts Avenue
Cambridge, MA 02139
mitpress.mit.edu

The MIT Press would like to thank the anonymous peer reviewers who provided comments on drafts of this book. The generous work of academic experts is essential for establishing the authority and quality of our publications. We acknowledge with gratitude the contributions of these otherwise uncredited readers.

This book was set in Stone Serif and Stone Sans by Westchester Publishing Services. Printed and bound in the United States of America.

Library of Congress Cataloging-in-Publication Data

Names: Myin, Erik author
Title: Of a different mind : an introduction to the philosophy of
mind and psychology / Erik Myin.
Description: Cambridge, Massachusetts : The MIT Press, [2026] |
Includes bibliographical references and index.
Identifiers: LCCN 2025028957 (print) | LCCN 2025028958 (ebook) |
ISBN 9780262052993 paperback | ISBN 9780262052986 pdf |
ISBN 9780262053006 epub
Subjects: LCSH: Philosophy of mind—Textbooks | Psychology—
Philosophy—Textbooks | LCGFT: Textbooks
Classification: LCC BD418.3 .M95 2026 (print) | LCC BD418.3 (ebook)
LC record available at https://lccn.loc.gov/2025028957
LC ebook record available at https://lccn.loc.gov/2025028958

EU Authorised Representative: Easy Access System Europe, Mustamäe tee 50, 10621 Tallinn, Estonia | Email: gpsr.requests@easproject.com

To Inez Germeys

Contents

Acknowledgments

> I get by with a little help from my friends.
>
> —The Beatles, *Sgt. Pepper's Lonely Hearts Club Band*

Writing this book has been on my mind for a long, long time. But life, in its whimsical ways, rarely allows plans to unfold just as we hope they might. Something out of the ordinary had to happen. COVID certainly took care of that. I don't want to begin these acknowledgments by giving credit to a highly lethal virus. But still, it was because of the COVID lockdowns that I found the time to start to engage in the very extensive reading that this project required.

If the pandemic was a catalyst, the process leading to this book stretches out long before it. In retrospect, it spans at least my whole academic career, and of all of the reading, thinking, and writing I have done during it has in one way or another contributed to it. I have relied on pieces I wrote for my master's thesis, for courses I assisted when I was a PhD student in Brussels, and much other material written along the way. Therefore, I should thank each and every person who has been involved in any way in my philosophical adventure. I single out my parents to thank them for making my philosophical trajectory possible at all, and for their support throughout. As regards naming other people, I can only go on to explicitly mention those people who were of importance in the coming about of this book.

In the first place I want to thank the MIT Press, and especially Philip Laughlin. Phil has shown interest and support for this project from the first moment I contacted him about it. He took great care in guiding and assisting the process from proposal to final manuscript. It is wonderful to see that in times when pirating academic works has become common practice,

Phil and the MIT Press still dedicate so much effort to a work destined to for print. Many thanks also to all the people that have been involved in the many stages of the production of this text. I cannot forget to express my gratitude to the reviewers of the proposal for this book, and later of a full version of the manuscript. Their comments, in both stages, have been of great value and help.

Even when my plans for this book were still vague and immature, I had the MIT Press in mind as the best outlet for this book. That is because I always thought of my project as related to Paul Churchland's *Matter and Consciousness*. Though I am not an eliminativist, I greatly admire Paul's thinking and writing, and I have fond memories of visiting him and Patricia in San Diego in 1993–1994.

I owe my sincere and eternal gratitude to the students for the course "Philosophical Psychology" at the University of Antwerp in 2023, 2024, and 2025. They were presented with earlier versions of this manuscript, rife with mistakes, omissions and typos, especially in 2023. If they served as guinea pigs, I hope no permanent damage has been done to any of them. They provided me with invaluable feedback, which allowed for countless improvements.

I also want to thank the doctoral students and postdocs that formed part of my research group during the writing of the manuscript. Apart from those of the group that will be mentioned further on, I thank Amit Gur, Luca Roccioletti, Zuzanna Rucinska, and Wouter Van Hooydonk for their comments during a Work in Progress meeting dedicated to this book. I am indebted to Thomas van Es, and Michiel Esseling for providing detailed and very useful feedback on specific parts of the manuscript. David Schoute and Ludger van Dijk offered very insightful comments after reading a complete version. So did Marco Facchin, Giula Di Rienzo, and Farid Zahnoun. In addition, I owe much gratitude to Marco, Giulia, and Farid for teaching "Philosophical Psychology" in 2024 and 2025. I am also sincerely grateful to Jan Van Eemeren for taking over other teaching tasks in 2023–2024.

My old friend Jan Timmermans, whom I seem to have known forever, is entitled to a paragraph of his own. He took the effort to read a complete version of the manuscript and gave me very welcome comments and heartening encouragement.

I want to express my sincere gratitude to my academic collaborators and colleagues, especially those I coauthored with. They taught me many things, including how to become a better writer. Of those I wrote pieces with, published or not, I single out: Daniel Hutto, J. Kevin O'Regan, Alva Noë, Louise Barrett, Malika Auvray, Michael Kirchhoff, Karim Zahidi, Kirsten Poortier, and Peter-Paul Verbeek. Special thanks also to Dan Hutto for forming the REC team and, together with Farah, for hosting me multiple times in Wollongong. A lot of this book was written during my stays Down Under. I learned much from the comments given on presentations of parts of the book in Wollongong from Glenda Satne, Lorena Sganzerla, Marco Degano, Mads Dengsø, Benedetta Cogo, Ian Robertson, Max Capuccio, and Shaun Gallagher. I keep fond memories of Stefan Delfgou and Joanna and their sons, for the wonderful company they provided in the Gong. Thanks to Tim Bayne, for taking the time to meet in Melbourne while I was working on the manuscript there, and for thoughtful exchange between fellow writers of introductions to the philosophy of mind.

My colleagues Fred Keijzer and Pierre Steiner deserve a special place. Their encouragement to actually carry out my plan for this book has been very significant to me. Besides, Fred and Pierre have always exerted an influence on me through their excellent work. Thanks to both, too, for meeting during my sabbatical dedicated to this project. Another colleague from Groningen that I visited then and whom I owe sincere gratitude is Rob Withagen. Rob was so kind to give me excellent comments on various versions of draft chapters and that allowed me to significantly improve them.

I can't name all the people I met at conferences and who contributed to many of the ideas presented in this work. I do want to mention Albert Newen, Tobias Schlicht, and Marcos Silva for organizing various events in Bochum and in Fortaleza, that provided ample opportunity to discuss work related to this book.

Closer to home, I have space to mention only a few of my colleagues. These are Joachim Leilich and Peter Reynaert for welcoming me in their research environment, Luc Van Liedekerke for replacing me as deputy head of department in 2023, and Geert Van Eekert for replacing me as head of department for some months in 2025. I owe thanks to Arthur Cools for repeated encouragement for this and other projects, and to Katrien Schaubroeck and Judith Martens to participate in the reading group on Anscombe's *Intention*. Walter De Mulder must be sincerely thanked, too.

Walter, linguist and former dean of the Faculty of Arts at my university, has always been interested in my work and found time in his busy schedule to read a version of the whole manuscript. His enthusiastic reaction and his many excellent comments were of great help for further refining the text. A very special thank you goes to Laura Butzen, who generously volunteered to read and correct the proofs of the manuscript during the final weeks. As I expected, she did a brilliant job.

For financial support by funding a sabbatical during part of 2023, I thank the Fund for Scientific Research Flanders (FWO) and the Research Council of the University of Antwerp. I also benefitted from support from the European Commission through a grant for the Marie-Curie Early Training Network REPAIRS and for the Marie Curie Postdoctoral Fellowship *Beyond the Information Paradigm: From Prediction to Anticipation*, given to Farid Zahnoun. Of the people that participated in the REPAIRS consortium, or in its activities, I can only mention some that directly influenced my work on this book. They are: Raoul Bongers, Frank Zaal, Gregor Schöner, Marijn Hafkamp, Reinoud Bootsma, Joanna Rączaszek-Leonardi, Heleen Reinders-Messelink, David Travieso, David Jacobs, Joanne Smith, Pedro Passos, Paula Siva, and Bill Warren.

During the preparation of this manuscript, I picked up sculpture classes again. I found attending these classes very important during the parts of the sabbatical period I was in Belgium. Sculpting allowed me to bring my mind to rest by changing its focus. Besides, sculpting is a form of thinking, related in many ways to writing. Learning about the one means learning about the other. I thank my teacher, Wouter Mulier, for his skillful guidance.

This brings me to another sculptor, Mark Manders. I am extremely grateful to Mark and to Yehbonne Bien at Galerie Xavier Hufkens for allowing me to use a rendering of his marvelous work *Vertical Clouds* for the cover of this book. Many thanks also go to my secondary school classmate and friend Werner Van Peer and his brother Dirk for drawing my attention to that piece of art in this context.

Finally, for broad support, intellectual, emotional, and everything in between, I thank my three daughters, Elise, Charlotte, and Laure, and my wife, Inez. It is always a delight to have them as company, in good times and in bad times. Inez also read a complete late version. She knows more about psychology and scientific research—not to mention life—than I ever will. Her unique perspective and sharp mind allowed me to improve this manuscript in countless ways. It is to her that I dedicate this book.

1 Introduction: Thinking About Minds Differently

In the beginning was the act.

—Johann Wolfgang von Goethe (1808/1962, p. 82)

Why write another introduction to the philosophy of mind and psychology, and why call it *Of a Different Mind*?

The answer to those questions is that much of the contemporary philosophical discourse about the mind—in textbooks and in professional philosophy and science—is dominated by a way of thinking and talking about the mind. A prominent picture is that of the mind as a naturally evolved kind of computer. On this view, evolution has equipped us with a biological information processor inside our heads, and it is thanks to its computational operations that we perceive, act intelligently, and think.

There is variety within views of the mind-as-computer, with important differences between them. But other views reject the idea that the mind is fundamentally an information processor or computer. Again, with many differences between them, these accounts tend to emphasize that one must understand the mind in action-based terms. The mind arises in and through action, and thinking comes later—it could even be a form of action.

In this book, I want to provide an overview of the philosophy of mind and philosophy of psychology in which these action-first positions are not seen as of secondary importance. There have been periods in the history of Western thought when they were the norm, and they are certainly experiencing a revival right now, under such labels as "embodied," "embedded," "extended," "enculturated," and "enactive" cognition.

Thinking about the mind is shaped by the history and context in which it occurs. No matter how abstract thinking can become, no philosopher operates in a void. Problems about the mind and proposed solutions are steps in the developing history of thinking. A prominent source of influence comes from science, including other sciences than those aimed explicitly at the mind. Because of the historical and intellectual embeddedness of philosophical thinking about the mind, I will pay considerable attention to the scientific revolution of the seventeenth century, the rise of experimental psychology, the emergence of evolutionary thinking in biology, and the so-called cognitive revolution of the second half of the twentieth century.

Readers familiar with existing contemporary introductions to the philosophy of mind will recognize the main protagonists, like dualism, behaviorism, the identity theory, functionalism, and eliminativism. But then, they will be dealt with differently here.

At the end of each chapter, including this one, a somewhat extended bibliographical note will be found. These passages are meant to be read, because they contain valuable additions to the main text. At the end of the book, a glossary contains a selection of important philosophical or technical terms.

Now on to the action.

Bibliographical Note for Chapter 1

A classical treatment of philosophy of mind is Kim (2011). A different perspective can be found in a book that was inspirational for this one: Paul Churchland's *Matter and Consciousness* (2013). A contemporary book, in the same intellectual tradition as Kim's, but with many innovations and contemporary additions is Bayne (2022).

One significant omission in this book is the scant attention given to the phenomenological tradition, of thinkers like Edmund Husserl, Maurice Merleau-Ponty, and many others. I have left this branch of philosophy mostly out for reasons of space and because there are two excellent treatments broadly in the spirit of the current volume available, namely Gallagher and Zahavi (2021) and Käufer and Chemero (2021). Käufer and Chemero also give attention to another thinker and another tradition missing from the picture painted here: William James and Gestalt psychology.

2 Mind Versus Body: Dualism and Its Reasons

If a mirror should possess a soul, it would see the image that is formed on it.
—Avicenna (1027/1955–1958, p. 60)

2.1 Dualism Versus Monism

When we say an athlete has recovered physically but not mentally, we distinguish between mental and physical recovery. But what is this difference we make between "physical" and "mental" precisely? Does the distinction point to some fundamental difference in reality? Is reality divided into two kinds of things and processes: those that are physical and those that are mental? Or does reality have two irreconcilable aspects? *Dualists* about the mental and the physical claim that mind and body differ in some fundamental, deep sense.

Why would one endorse dualism about mind and body? There are possible religious reasons to be a dualist. Many religions hold there is an afterlife in which persons, after their earthly decease, live on in some way. But if the body is dead, burned, or decaying, what lives the afterlife? Many religions invoke the soul or a related concept: a purely mental entity that is, during life, conjoined with the body but leaves it upon earthly death to continue to exist on its own elsewhere.

However, one must not necessarily resort to religious beliefs to conclude that mind/body dualism is correct. To many, the fact that we think thoughts and enjoy experiences shows us that these thoughts and experiences are unique and different from what happens in the material world. One can hold that thoughts are obviously different from the physical. After all, one can think about a mountain at any time, in any place, and one could do so after the mountain has ceased to exist, or even when one finds out the mountain has never existed. Thoughts don't seem to have the "heaviness" of material things; they escape the confines of space and time in ways material objects cannot. Or perhaps it is not in thought but in sensory experience that we directly meet the mind. Could there be a more glaring difference than between your smell or taste of a strawberry and the physical processes that are somehow involved in your smelling or tasting? The strawberry is but a collection of molecules, themselves collections of atoms. So are your body, mouth, tongue, and brain. But where do you find the fragrance or taste of the strawberry in the swirls of atoms? To account for the fact that we consciously experience the smell and taste, must one not assume that the objective processes create or are accompanied by something of a very different nature—something subjective that brings in the "feel"?

Throughout the history of Western thinking about the mind, mind/body dualism has been the default option (Reed, 1997). But that doesn't mean it has been uncontested. From ancient times on, some thinkers have endorsed *monism*, the view that, despite differences, ultimately, what's mental and what's physical are of one kind. "Materialism" or "physicalism" are names for monisms, according to which, at the bottom, the mental is material. To hold that experiencing or thinking are brain processes and nothing more is one way to be a monist of physicalist or materialist stripe.

One also finds a different kind of monist breed. According to mentalist monisms, matter is fundamentally mental. This position is called "idealism," for it construes the material world as made up of experiences and ideas. Strictly speaking, so says idealist monism, mountains are not out there. Instead, mountains exist in conscious beings' experiences and thoughts about mountains. To be is to be perceived, or so George Berkeley (1685–1753), the famous British empiricist, proclaimed, based on the view that, in the final analysis, material objects are made up of ideas in the mind about material objects (Berkeley, 1710).

Yet, mind/body dualism has been dominant in the history of Western thinking, at least until relatively recently. Nowadays, physicalism, or materialist monism, especially in the Western academic milieu, has become more prominent. This doesn't mean every contemporary philosopher of mind is a physicalist. Physicalism remains contested. Interestingly, many contemporary philosophers who are skeptical of physicalism and lean toward dualism cite experience as the source for their position. They think physicalism is unfit to explain tastes, smells, and other subjective experiences. Historically, many defenders of mind/body dualism in Western intellectual history have resorted to the unique nature of thinking to make the case for dualism.

This chapter will present some prominent historical dualist views. Engaging with the dualism of Plato, Aristotle, and Descartes remains worthwhile because, even if their varieties of dualism are no longer accepted, the reasons they offered for their views still play a role in current theories about the mind. To see thinking as what makes minds into minds remains a powerful idea. As said, contemporary doubts about physicalism focus on sensory experience as something unexplainable by a materialist view. But, as this chapter will make clear, the idea that conscious sensory experiences are opposed to objective processes might owe much of its appeal to a still impactful view of nature that motivated the dualism of science-minded modern thinkers like René Descartes.

2.2 Premodern Dualism

2.2.1 Lasting Greek Foundations

Perhaps unlike religion-based dualism, the dualism of philosophers is meant to be argued for. In the hands of philosophers, the truth of dualism becomes something to be rationally proved, or at least, sustained with arguments. They think the truth of dualism can be established by reasoning. In Plato (428/427–348/347 BCE), one finds such philosophical arguments in support of dualism (on Plato, see, e.g., Adamson, 2014, part II). Plato's plea for mind as distinct from matter forms part of his overall philosophy, dominated by a more encompassing dualism between form and matter. The world of matter, according to Plato's well-known and extremely influential views, is a world of particular things, such as wheels or trees. These particular things are temporal and perishable. The world of forms contains immutable and unperishable general ideas, such as the idea of a circle or of a tree. Matter can instantiate ideas. It is by instantiating the general idea of circularity, for example, that a particular wheel is circular. Plato argued that only by accepting this overall dualism, one could explain the existence of distinct particular things with the same properties and the possibility for us to have knowledge of such things and properties. One knows of the wheel that it is circular, in general: What one knows about is circularity, as it occurs not only in this wheel, but in all circular things. Therefore, so reasons Plato, one gets in touch with forms. But, he continued, for a mind to be able to get in touch with forms, it must be of the same nature as those forms. That is, the mind must be immutable and eternal itself. As the body is perishable, the mind must be a distinct entity, conjoined to the body in earthly life but able to exist on its own. Here we have Plato's mind/body dualism, a dualism as clear-cut as possible. Embedded as it is in the wider dualism of form and matter, the fortunes of Plato's dualism about the mind depend on the fortunes of his broader dualism. Even if correct, the wider dualism has some problematic implications for Plato's dualism about the mind. For example, forms are general, and if the mind is of the same nature as forms, it too must be general. But you have your mind, and I have mine; both seem to be particular minds. Now, how can mind be both particular and general? Why is there not one general mind instead of the multitude of minds that exist among each other? So even if there were no philosophical objections to Plato's overall dualism, his dualism about the mental would face tough questions to answer. But there are very powerful arguments against his overall dualism, many of which were formulated already by his pupil Aristotle (on Aristotle, see, e.g., Adamson, 2014, part III).

Aristotle (384–322 BCE) said he could not but reject the thinking of his esteemed master, because he loved Plato, but he loved truth more. Aristotle objected to the core of the overall dualism: the insistence on an absolute distinction between form and matter. If such distinction is indeed as required by Plato's theory, how can the abstract form (circularity) be *in* the wheel, instead of leading a separate existence, in Plato's other world? Aristotle thought Plato had made a mistake by construing properties of things—the roundness of the wheel—as things that existed in and of themselves. Instead, circularity, so held Aristotle, exists only in particular things, namely in all circular things. Following Aristotle, one can criticize Plato for treating what's general (circularity) as if it were a particular itself (the form circularity).

So, in Aristotle's view, general properties are *in* things: circularity in this wheel, and again in that disc. He did make a distinction between matter and form, however, just as Plato did. Matter, so thought Aristotle, was cut up in the various individual entities that exist. But matter in itself is indeterminate. It needs to be shaped by forms. Forms inhere in pieces of matter and thereby make them into bearers of properties. So, though forms in general don't exist independently of the matter into which they inhere, they shape matter, which would remain incomplete without them. In his work known as *De Anima*, Aristotle considers the mind to be the form of the body of living organisms (Aristotle 355 BCE/2011). One way to interpret this is to think of the mental as the way the living body is organized: the way its parts interact with each other and collectively with the environment to give rise to bodily and mental activities. Such an organizational view of the mind has never entirely left philosophical thinking about the mind.

In fact, Aristotle distinguished three souls, or what we would consider as "mental" principles shaping the body. He distinguished among a vegetative, sensitive, and intellectual soul. Natural desires, such as for food, survival, and procreation, derive from the vegetative soul. Perceiving involves the sensitive soul, and thinking the intellectual soul. Animals share vegetative and sensitive souls with humans, but only humans have an intellectual soul. Moreover, the intellectual soul is of a different nature than the vegetative and intellectual souls.

Yet, though it looks like Aristotle's conception of mind as form (in the specific Aristotelian sense) allows it to be construed in a monistic way, he did take a dualist turn in arguing for the unique status of the human mind. Again, as in Plato, the basis for the dualistic view of the mind lay in knowledge and the general nature of knowledge. Aristotle argued that, as the mind can know so many things (yellow, sweet, circular, big, and so on), it must be itself general and not particular. Aristotle thought of both perception and knowledge in terms of the receiving, by the perceiver or knower, of the forms inherent in the perceived or known objects. To see yellow is to perceptually receive the form yellow. To think yellow is to intellectually unite with the form yellow. Our perceptual senses of seeing, hearing, feeling, smelling, and tasting are subserved by sensory organs that are fit to receive their particular kinds of forms. For example, the eye is apt to receive color forms but not auditory forms, which fall in the ear's domain. Unlike the specific sensory organs, the intellect can unite with any form, because everything can be thought. Therefore, unlike perception, the intellect has no organ, or so says Aristotle in *De Anima*, and therefore, it is not material. That is, and ultimately agreeing with Plato, it is because of its general capacity to think abstract thoughts that the mind is fundamentally different from the body.

Again, the roots of a dualistic conception of the mind lie in the abstract nature of thought-like activities (including here perception). By following Plato in regarding the abstract nature of knowledge (including perception) as a reason for dualism about the mental, Aristotle inherited some of the problems with which Plato's view was faced. For example, as was widely discussed by Arab followers of Aristotle in the Middle Ages, how can there be separate intelligences in you, me, and other people, if the intellect is general? Some of Aristotle's Arab followers, spearheaded by Averroes, did bite the bullet, and accepted that strictly speaking, there was in the entire universe but one encompassing intellect.

2.2.2 Mind/Body Tensions in Medieval Thought

Averroes, or Ibn Rushd (1126–1198), was a medieval thinker working in the Islamic tradition in what was then the Islamic Andalusian peninsula (now Spain). He was a commentator on and staunch defender of Aristotle, whose work he tried to further develop. This led him to argue, on Aristotelian principles, for the existence of one universal mind as what underlies all thinking. Every individual thinks only by participating in the activities of this one mind. As Averroes pointed out, the conclusion that at bottom there is only one mind follows from what Aristotle wrote about the intellect. If the intellect engages with the abstract, and knows by forming a union with it, then the mind itself is abstract and not individual. Hence, there is only one single mind (Adamson, 2016, p. 187). But how to make sense of the feeling that we have individual minds and can think individual thoughts? To account for individual thoughts, Averroes invoked a lower kind of cognitive activity that could accompany the having of abstract thoughts. Adamson characterizes such "cogitation" as not proper, universal understanding but an allegedly particular process of "active consideration of particular things we have seen or otherwise experienced" (Adamson, 2016, 192).

A prominent Christian medieval philosopher was Thomas Aquinas, from Aquino in Italy (1225–1274). He was as committed to Aristotle as was Averroes but sought to harmonize the teachings of his Greek master with Christian theology (on Thomas Aquinas, see, e.g., Adamson, 2019, especially chapters 36 and 37). Aquinas rejected the "single mind" view of Averroes as being incompatible with the immortality of individual souls, a cornerstone of Christianity, as of many other religions. He defended the immortality of individual intellectual souls of human beings. He followed Aristotle's tripartite division of vegetative, sensitive, and intellectual souls—understood as forms of the body. He also accepted Aristotle's grounding of mind/body dualism in the intellectual soul's capacity to think abstract thoughts. Thomas warned for taking the three different souls to be three different entities. Instead, so he stated, they form different aspects of one single soul. Yet only the human soul has the intellectual aspect, and only the human soul is immortal.

Only the aspect view is able, so he argued, to do justice to the unity of mental life, for example, the way thinking and perception are intertwined. He recognized that body and mind were entangled and claimed that all thinking makes use of sensory elements, so that the intellect depends upon sensitive capacities and, therefore, on the sensing body.

When people die, their immortal souls continue to exist, but now without being sensorily affected. The soul then finds itself in an unnatural condition. When the Day of Judgment arrives, souls are reunited with their resurrected bodies, and the soul returns to its natural mode of existence to enjoy perpetual happiness in heaven or eternal damnation in hell. Throughout, and despite its explicit dualism, Thomas's approach to the mind recognized an essential connectedness of mind and body. Indeed, an early exponent of contemporary "embodied cognition" (see chapter 10), neuroscientist Walter Freeman, mentions Aquinas as a forerunner (Freeman, 2009).

2.3 Cartesian Dualism

2.3.1 The Mechanization of the World Picture

Roughly until the seventeenth century, views based on Aristotle remained dominant in Western thinking. The overall view of nature, of the universe, and of biological and physical processes was in line with the writings of Aristotle. It would be mistaken to attribute this dominance to an excessively deferential attitude, for Aristotle's writings contain elaborate and coherent views of the universe as a whole and many processes occurring in the physical and biological realm. As the historian of science David Lindberg writes, "Aristotle prevailed through persuasion, not coercion" (Lindberg, 1992, 68).

There had, of course, always been thinkers who contested Aristotelian theories and claims. Alternative views of nature and of specific processes had been offered. It was only in the seventeenth century, however, that the balance of power started to shift. Thinkers challenged both the Aristotelian philosophy of nature and its accompanying astronomy, biology, and physics. A major target of criticism became the Aristotelian teleological view of natural processes. As discussed, Aristotle saw natural entities—including objects and organisms—as matter shaped by forms. Forms not only gave shape to matter, but the forms inherent in objects and organisms also determined how these objects and organisms moved and acted. Forms were thought of as inner causes that, by their influence, led matter to specific goals. Therefore, Aristotelian explanations for all natural processes were teleological. Natural processes, such as biological growth or physical motion were seen in terms of inherent goals. Just like the growth of the seed of a plant seemed to follow a definite plan, so the fall of an object released from a hand was analyzed and explained in terms of an inherent tendency of heavy objects to be near to the center of the earth (Lindberg, 1992, chapter 3).

Seventeenth-century philosophical literature is replete with attacks on the Aristotelian view of natural objects as animated by forms. The teleological view of natural entities became criticized as being anthropomorphic, that is, as modeling objects on persons and as naively taking such models for real. Form-based explanations of natural processes drew sarcasm for being empty because the inherent forms seemed miraculously to contain all that was needed to explain the movements of the entity (Shapin, 1996, p. 30). Thus, British philosopher Thomas Hobbes (1588–1679) ridiculed the Aristotelian explanation of the movement of an object to the earth by invoking the influence of the form heaviness in a body as follows:

> But if you ask what they mean by *heaviness*, they will define it to be an endeavor to go to the center of the earth, so that the cause why things sink downward is an endeavor to be below (which is as much as to say that bodies descend or ascend because they do . . . as if stones and metals had a desire or could discern the place they would be at, as man does. (Hobbes, 1651/1994, part IV, chapter 46, 462)

Thomas Hobbes was but one of many thinkers who was actively involved in what has been called "the mechanization of the world picture" (Dijksterhuis, 1950/1961). Instead of being driven by inherent goals, nature and natural processes became modeled on mechanical machines. The metaphor of the universe as a giant clockwork became ubiquitous (Shapin, 1996, 31–37).

Mechanical systems such as clockworks consist of a number of parts that operate in a fixed and predictable way. Understanding how the different parts of a mechanism work and how they are put together allows one to understand the workings of the complete mechanism or the whole machine. The movements of the hands of the clock no longer look mysterious once one grasps how they are produced by the simple motions of the wheels and springs inside the clock.

In the mechanistic worldview, nature was seen as a giant mechanism, the physical workings of which could be understood by referring to the basic operations of its components. Moreover, the workings of a mechanism and its components are measurable, and it is possible to give a precise mathematical treatment of how the operation of one component changing over time affects the operation of another component—for example, of how movements of the gears and other parts within the clock make the hands of the clock move over its face.

The mechanization of the world picture was also a mathematization of the world picture. A science such as physics was a science that allowed us to describe mathematical regularities of motion and other natural processes. Hence, physicists like Galileo Galilei (1564–1642) and Isaac Newton (1642–1727), who were able to describe the laws of motion, acquired immense intellectual prestige. This fate befell, in particular, Newton, who was able to provide a unified treatment of both the motions of objects on earth and the orbits in our solar system. With no more than three physical laws, Newton was able to describe the path of cannonballs, the tides of the seas, the motion of the moon around the earth, and the motion of the earth with respect to the sun (Newton, 1687/1999).

It would be a mistake to think that the mechanization of nature led to an overall attitude of hard-nosed scientism: the belief that the answer to all serious questions lies in science. Such scientism might leave no place for religion. It might also be thought that the purging of inherent goal-imposing forms from nature might have led thinkers to expel nonphysical minds from bodies. But, apart from some exceptions perhaps, none of this happened. On the contrary, mechanistic philosophers cited their view of nature as providing compelling reasons for religious faith and for the existence, at least in humans, of an immaterial, nonmechanistic soul (Shapin, 1996, part 3). Mechanistic thinkers objected to Aristotelian forms as explanatory principles but not to what these forms were meant to explain. Even if mechanistic thinkers saw nature, often including animals, as machines, they did not think of nature as entirely devoid of goals. Seeing nature as a clock implies a designer of the clock and thus led them to God as the creator and designer of the universe and all it contains. Goal-directedness was no longer inscribed into nature from the inside by Aristotelian forms, but by the precise arrangement of its mechanical components by a supernatural external designer. Mechanistic thinkers found goal-directness also within themselves, in their conscious thinking and willing. Again, with only a few exceptions, the mechanistic thinkers did not think of their own thinking and willing as mechanical or clocklike, but rather as something obviously goal-directed by itself. Thinking looked undeniably nonmechanical and, therefore, seemed to provide incontrovertible evidence for the existence, in themselves and fellow humans, of a nonphysical mind (Shapin 1996, parts 1 and 3). That a mechanistic view of nature and dualism about minds and bodies were mutually supportive is abundantly visible in the writings of René Descartes. Descartes was at the same time a radical mechanicist and radical dualist.

Before we discuss Descartes's dualism, one other aspect of the mechanization of the world picture must be discussed: the emergence of a distinction between primary and secondary qualities. The mechanistic view of nature brought a profound change in how sensory experiences of properties like taste and color were seen. To grasp this, consider the Aristotelian account of color perception. Aristotle and his followers saw perception as a sharing between perceiver and object of the forms inherent in material objects. An ochre ball of clay is what it is because it incorporates the forms ochre and round. A perceiver can see that ochre and round piece of clay because they possess organs that themselves can receive the forms ochre and round. When the perceiver turns their eyes on the piece of clay, the forms ochre and round become shared between and inform—give shape to—both the perceiver and the object.

When mechanistic philosophers chased forms out of nature, this view of perception was no longer possible. Perception could no longer be seen as a communion of forms. As a result, some of the properties that we sensorily experience, such as colors and tastes, were assigned a special status in mechanistic philosophy. The key distinction underlying this special status is often traced back to one of the fathers of the mechanistic worldview, Galileo Galilei. It is a distinction between properties he calls "primary" or "real" and properties that belong only to consciousness. Primary qualities are those that can figure in mechanistic explanations. Examples are form, motion, or number. Indeed, it are such properties one would refer to if one gave a mechanistic explanation of the speed and direction a ball was moving in after having been hit by another ball. The mechanical properties are measurable: They can be assigned a number and treated mathematically.

In his book *The Assayer*, written in 1623, Galileo Galilei introduces the distinction by considering that primary qualities are *in* the physical objects. They are inseparable from the objects, in the sense that if we think of an object, we have to think of its figure, size, position, its motion or rest, and its number (Galilei, 1623/1957). However, next to these primary properties, there is another kind of properties that can show up in the experience of an object, whose characteristic is that they do not seem to inhere in the object but rather in the experience of the object. Galileo wrote:

> I think that tastes, odors, colors, and so on are no more than mere names so far as the object in which we locate them are concerned, and that they reside only in the consciousness. Hence if the living creature were removed, all these qualities would be wiped away and annihilated. (Galilei, 1623/1957, 274)

To get a grip on the distinction, think about color and its experience. The colors that we experience, such as red, green, and purple, correspond to different physical properties of light. But, as Isaac Newton famously stated about physical light, in a passage underscoring Galileo's point: "For the Rays properly speaking are not coloured" (Newton, 1730/1952, Bk. I, Pt. II, Prop. II, Thm. II, Definition, pp. 124–125). What Newton means is that the physical differences between the light that is seen as green versus the light that is seen as red give us no clue about the felt differences. The rays might be composed of broader or narrower waves, but those physical differences are different from the felt difference between red and green. The felt differences belong not to the rays but to the perceivers. They concern "qualities," the different way experiencing red feels from how experiencing green feels. Unlike a quantitative property like length, a qualitative property can't be fully characterized by a number. There's nothing more to say about length once one has specified the length with a number. But even if you say that you hear the flute twice as loud as you heard it before, it seems you have not fully characterized the experience of hearing a flute. To get a grasp on the experience, you need to point to what occurs when you actually hear a flute. Resonating with what Galileo and Newton wrote, qualitative properties are then taken to exist only when perceivers have sensory experiences. There are other qualitative properties aside from color and sound. What goes for color and sound seems to go for taste and smell. The shape of an orange or the number of apples on a tray is objective, out there and measurable. The form doesn't change or come into existence when the orange is perceived or when the apples on the tray are counted. Yet it makes sense to say that the taste of these fruits exists only when they are eaten. From the writings of John Locke (1632–1704), this distinction between quantitative and qualitative properties would be known and widely referred to as the distinction between "primary" and "secondary" qualities (Galileo only mentioned "primary").

2.3.2 Descartes's Radical Mechanicism

The view that properties like color and taste belong to conscious subjects and not to the objective, physical world can lead to mind/body dualism. If color or taste experiences are radically unlike physical properties, so one can reason, the mind must be nonphysical. Such reasoning underlies some prominent defenses of contemporary nonphysicalistic views of the mind. These will be discussed in chapter 9.

But although the mechanistic worldview of the seventeenth century formed a fertile ground for justifying mind/body dualism by referring to sensory experience, it was not the preferred way to ground dualism in that period. To be sure, mind/body dualism reigned, but it was not argumentatively motivated by considering sensory experience. Rather the justification of dualism retained the focus on thinking, in particular, a pure form of thinking that was crucial for ancient and medieval dualisms. In the work of René Descartes, this connection between thinking and dualism became very explicit.

French-born René Descartes (1596–1650) was a foremost mechanistic thinker. He was a mathematician, a scientist mastering several disciplines, and a philosopher. He innovated mathematics, and we still call the method of representing space by a number of orthogonal lines a Cartesian coordinate system. He was a convinced mechanicist and critic of the Aristotelian form-based view of nature (Dijksterhuis, 1950/1961, part IV, III).

Descartes promoted a particularly austere form of mechanicism. In this view, all natural causation was modeled on the paradigm of one physical object altering the motion of another through direct contact. A moving ball striking a stationary one exemplifies the key features of this picture of natural phenomena. Physical objects C and E, both "bodies" or particles, come together in space and time, and because of contacting E, object C causes a change in E (as well as possibly in itself). The movement of E is caused not by an inner form or nature but by the movement of C. Descartes devised imaginative mechanistic models for many physical phenomena, including the motions of the planets, magnetism, and the reflection and refraction of light.

Descartes's zeal for mechanistic modeling did not stop at phenomena many—then and now—would consider "mental." He devised a mechanistic model for the pain reflex, as when one withdraws one's foot from a flame. Fast-moving fire particles were imagined by Descartes to move the skin of the foot so that it pulled a nerve going from the foot to the brain. Pulling the nerve opened up gates through which minuscule particles could start to flow to effect both withdrawal of the foot and the orientation of the head and eyes to look at the flame (AT XI, 141; CSM 1, 101—see the references section under Descartes for explication of the abbreviations).

Descartes went far beyond reflexes, to elucidate the bodily aspects of perception, memory, and imagination. It had been customary, in ancient and medieval traditions, to think of perception, memory, and imagination as involving some kind of "image" or "phantasma." These images, conceived of as inner entities, were formed when perceiving things, could be stored for later retention by memory, and could be produced at will by the imagination. In Aristotle's system, animals and humans could enjoy such capacities because of the vegetative and sensitive soul (Connell, 2021). With characteristic mechanistic fervor, Descartes declared the vegetative and sensitive soul to be unnecessary to account for basic capacities for perception, memory, and imagination (AT XI:202, CSM I:108). He thought of images as physical entities or patterns in the brain created by processes like pushing and pressing (AT XI:177–178, CSM I:106–107). These patterns were formed in perception, could be recreated in remembering, and created in imagination.

Just like it would be wrong to conclude that the overall mechanization of the world picture led to endorsing a physicalist view of reality, it would be wrong to conclude that Descartes's ideas on reflexes, perception, memory, and imagination led him to a physicalist view of mind. Descartes's mechanistic models were only one side of his picture of the mind. The reflexes and the mental capacities he had mechanistically modeled did not offer the complete picture of the mind, or at least not of the human mind. For human beings, the full story of the mind and its capacities required invoking something of an entirely different nature: the human soul. Humans can be conscious of pain; they can consciously perceive, remember, and imagine. But such consciousness requires that the mechanical events specified in the models also involve something fundamentally different: the nonphysical soul or mind. Only when physical events occur in concert with essentially nonphysical events does consciousness occur, or so Descartes claims, in line with his mind/body dualism.

But what, according to Descartes, is this nonphysical mind, and why should we believe in it?

2.3.3 Descartes's Radical Dualism

To Descartes, a closer consideration of machines and animal behavior provided him with one major reason for his mind/body dualism. He started by referring to machines. When he did so, he probably had in mind devices which he described in this way: "Clocks, artificial fountains, mills, and other such machines which, although only man-made, have the power to move of their own accord" (AT XI, 120; CSM 1, 99). More specifically he could think of the many artificial human- and animallike devices that were present in palace gardens that sprayed guests with water, played instruments, or spoke. (Cobb, 2020, p. 34; Riskin, 2016). However impressive such machines already were, upon reflection, Descartes found them to be limited in ways humans are not. He invoked language use to characterize the way in which human cognitive functioning fundamentally differed from how mechanical devices operate. So he wrote about machines:

> They could never use words, or put together other signs, as we do in order to declare our thoughts to others. For we can certainly conceive of a machine so constructed that it utters words, and even utters words which correspond to bodily actions causing a change in its organs (e.g. if you touch it in one spot it asks what you want of it, if you touch it in another it cries out that you are hurting it, and so on). But it is not conceivable that such a machine should produce different arrangements of words so as to give an appropriately meaningful answer to whatever is said in its presence, as the dullest of men can do. (AT VI, 56–57; CSM 1, 140)

The point made here by Descartes is that machines, being mechanical, act in lawful ways. They can be made to do certain things in certain circumstances. But humans are able to speak in all circumstances, even unforeseeable ones. In brief, mechanisms are, by their very nature, tied to function in certain conditions, but humans can function in any condition. In going beyond the specific circumstance-bound ways in which machines function, human cognitive functioning shows itself to be mechanistically unexplainable. No mechanical causal patterns can be found that cover the richness of human cognition.

Descartes made it explicitly clear that the fundamental distinction between humans and machines lies in the uniquely human capabilities for abstraction and universality, whereas machines remain bound to the realm of particularities. He wrote:

> Such machines might do some things as well as we do them, or perhaps even better, they would inevitably fail in others, which would reveal that they were acting

> not through understanding but only from the disposition of their organs. For whereas reason is a universal instrument which can be used in all kinds of situations, these organs need some particular disposition for each particular action. (AT VI, 57; CSM 1 140)

Descartes invoked a very similar line of reasoning to argue that animals, in contrast to humans, have no thinking minds. He offered the same assessment of animal behavior as he did of machine functioning. He admitted that animals can exhibit and be trained to exhibit many behaviors that align effectively with their circumstances. Like machines, they can show skillful performance in some areas. But he contended that such behaviors are always context-specific: They occur in this or that context and are entirely adapted to it. So, in such cases, only mechanical processes are at play, or as he writes, "it is nature which acts in them according to the disposition of their organs" (AT VI, 59, CSM 1, 141). Humans, on the other hand, can show intelligent behavior in any context. Humans therefore display rationality, which is an ability for thinking that transcends particular circumstances. In other words, human manifestations of intelligence bear the mark of generality, or are fit for all circumstances. They are abstract, while animal behavior remains concrete. We see that at the heart of Descartes's reasoning stands what was crucial in ancient dualisms: the abstract nature of human thinking.

For Descartes, it was just a matter of time before all of animal behavior would be understood completely in terms of the mechanistic science of nature. But mechanistic explanations, however appropriate for material motion, are, according to Descartes unable to account for truly intelligent action. For Descartes, this shows that human behavior has to be explained, not as lying along a mechanical causal chain beginning with material triggers in the environment but by invoking something entirely different, something that was not part of the mechanical order of nature. Tellingly, Descartes's partial mechanization of capacities for pain, perception and memory led him to the limits of mechanicism, and these limits appeared when thinking got involved.

But Descartes did much more to argue for the special nature of the human mind than contrasting human minds with automata and animals. Descartes also theorized the difference between mind and matter without resorting to facts about behavior. He thought that he could offer something like a mathematical proof of mind/body duality, just by investigating his ideas of thinking and of matter.

In his view, mathematics dealt with what he called pure ideas. Pure ideas are abstract. They are devoid of sensory aspects. Descartes illustrates the pureness he has in mind by considering the idea of a chiliagon: a thousand-sided figure. Descartes claims that it is only by pure thinking that one can conceive accurately of a chiliagon. One can imagine a chiliagon by calling to mind how it would look. Some image might come to mind, but one would never be sure it was a chiliagon and not a shape with a slightly different number of sides. Descartes uses the example of the chiliagon to support the claim that pure thinking is superior to imagination, which always relies on sensory aspects (AT VII, 72; CSM 2, 50–51).

Pure thinking, if performed well, leads to certainty. From considering the pure idea of a circle, one can conclude perspicuously that all of the points on its circumference lie at the same distance from its center. From pure ideas such as line, angle, triangle, and number, one can conclude, with absolute confidence that the angles of a triangle added together make up 180 degrees. Mathematical truths can be arrived at just by inspecting and reasoning from pure mathematical ideas.

Descartes then turned to apply the method of inspecting and reasoning from pure ideas to arrive at certainty, to his idea of mind. He wanted to start this investigation from something that was absolutely certain about the mind. But what could be such certainty about the mind? Descartes proposed that such certainty about the mind should be thought of as something the existence of which could not be doubted. Or, it should be something that remained true, even when everything about the mind came under doubt. Descartes pointed out that there was one such doubt-free property of the mind, namely, thinking. When questioning the existence of the mind, one could try to question anything, but one thing would withstand the assault of doubt: that one is thinking. For to doubt *is* to think. So, it is impossible to deny the existence of thinking when one is doubting, even when one is doubting thinking. With this reasoning, Descartes had found his desired certainty.

Descartes then continued to reason from the absolute certainty that a thinker's mind exists, to a fundamental difference between mind and bodies. To arrive at that conclusion, he investigated what else the idea of a thinking something contained except the very idea of thinking. His answer was straightforward: nothing at all. He came to this answer because he was convinced that he could conceive of something that had only one property: that it was thinking. Descartes claimed he could form the idea of bare thinking occurring without the existence of anything else. Conceiving this thinking-only possibility, so Descartes claimed, was possible without assuming there was any physical object. Indeed, Descartes claimed he could conceive of a thinking thing, not only in the absence of objects but in the absence of space. Thinking, therefore, is the only property that needs to be supposed for there to be thinking. On this line of reasoning, thinking was the one essential property, or in Descartes's language, the "principal attribute" of mind.

Then Descartes turned to investigate the idea of matter. Descartes asked, What property does one need to always invoke when one forms the idea of matter? He answered that it was extension, or taking up space. This is the one property that is always present when there is matter. Other properties of matter, such as motion, are only optionally present. Moreover, such properties presuppose extension. Only what is extended in space can move in space. Therefore, extension, but not motion, is the principal attribute of matter.

Descartes further reasoned that if one did not need to invoke the essential property of matter in conceiving of thinking and vice versa, thinking and matter must be different ways of existence, each characterized by its own self-standing principal attribute. On the one hand there is the substance, or most general kind of what exists, that he calls, in Latin, *res cogitans*—thinking thing; on the other hand, there is the substance he calls *res extensa*—extended thing, or mind versus matter.

Pure thinking provides Descartes both with a method and with the essence of mind. It is by engaging in pure thinking that Descartes can prove with certainty that mind and body are distinct kinds of things. Just by analyzing the ideas he has about the thinking mind on the one hand and material objects on the other hand, he makes a very strong claim, not only about his ideas of mind and matter, but about mind and matter in reality. The distinction he can draw between his ideas corresponds to distinctions in reality—Descartes calls these "real distinctions" (Rozemond, 2011). The arguments for mind/body dualism that referred to animals and automata were based on observations and reflections of actual phenomena: the behavior of animals and automata. In that sense they involved investigations of what happens in the world out there. Knowledge that is gained by observing things and events is called "a posteriori" knowledge, or knowledge that comes *after* observation. A lot of what is known about the world in the sciences is a posteriori in this sense. People have had to find out, for example, that the orbit of the earth around the sun is not a circle but more like an ellipse. In contrast, Descartes's philosophical argument for mind/body dualism is purely "a priori," or arrived at by thinking alone. Descartes concludes that mind and body are distinct because he can, with perfect clarity, separate in his mind the ideas of something mental—thought—and of something material—a physical object. Descartes arrives at his conclusion without doing tests and observing their results. One might try to find out if a lump of gold was pure or contained a bit of copper by heating it or by seeing to what extent it can be bent. Then one would arrive at one's knowledge of the lump's constitution a posteriori. But this is not how Descartes goes about in his philosophical investigation of the ultimate constitution of reality. His conclusion is arrived at entirely a priori. He doesn't manipulate matter to conclude that it is not thinking; he just proceeds by analyzing his idea of matter and by contrasting it with his idea of minds.

Descartes was confident that his mind had the required powers for reaching certainty about real distinctions by a priori reasoning and thus for finding out fundamental truths about the world just by entertaining and reasoning from clear and distinct ideas. After all, his mathematical work was performed in this way, and it seemed to provide conclusions applicable to the real world. Descartes also attempted to give argumentative justification for why his method worked (Wilson, 1978, chapter I, part 7). There he invoked the idea that the method was guaranteed because if his reasoning did not have these powers, the God that gave him his reasoning powers would have been a deceiver—and the idea of a deceiving God is in tension with the very idea of God. For all the argumentative brilliance that such proofs of the validity of reasoning might have in Descartes's hands, they inevitably look like attempts to lift oneself up by pulling on one's own hair. For only if one knew already that the reasoning was sound, one would be convinced by the proofs it provided. If one had no decisive answer about whether the reasoning was sound, why would one believe a proof was sound if it is constructed by that very reasoning?

2.3.4 The Embodied Mind, Cartesian Style

For Descartes, the mind can be self-sufficient and can operate on its own. That happens, for example, when it contemplates pure mathematical ideas. But often, so thought Descartes, human mental phenomena are of a mixed nature: Both the body and the mind are involved. In human perception, the mind takes up sensory elements that are caused by the body and envelops them in thinking, and this also happens in other human mental capacities, such as imagination.

To make possible the mental lives humans enjoy, body and mind have to come together. In perception, bodily stimulation has to meet the thinking mind. There must be a path from a body being affected to something like a conscious experience of what affects the body. In laying out this path, Descartes relied on a scheme that has been retained in mainstream views about sensation, perception, and thinking until now (Reed, 1982).

Descartes agreed with Galileo and many others on the distinction between primary and secondary qualities. He thought colors, tastes, and other secondary qualities, are not in the objects that affect our sensory systems. Instead, certain primary qualities in objects cause specific events in the sensory organs and nervous system. When these occur we experience secondary qualities:

> It must certainly be concluded regarding those things which, in external objects, we call by the names of light, color, odor, taste, sound, heat, cold, and of other tactile qualities . . . are not . . . anything other than various arrangements of the size, figure, and motions of the parts of these objects which make it possible for our nerves to move in various ways, and to excite in our soul all the various feelings which they produce there. (AT VIIIA, 322; CSM 1, 285)

Descartes further held that we know the objective world, and hence the primary qualities, based on awareness of secondary qualities (Reed, 1982). How this works can be seen by considering Descartes's account of our perception of the distance from an object we see to us. Roughly, Descartes theorizes that we see the distance of an object upon which both eyes focus or converge, in the same way a blind man holding two sticks can sense the distance of something at the apex of the triangle formed by the two sticks. If you consider the triangle formed between the apex (E), and the two places where the hands hold the stick, the blind man can know at what distance the apex is by knowing in which angles he holds the sticks. It is a basic truth of trigonometry that the distance is fixed by the two angles.

But in the same way, the mind can know the distance of some object based on the angle at which the eyes turn to bring the object into focus. This angle can be known by a tactile feeling corresponding to how much the eye is turned (Reed, 1982).

Descartes's overall scheme for how human perception happens has three stages (Reed, 1982). At the first stage, the body is affected in a certain way.

Figure 2.1
Blind man holding sticks, from Descartes's *Optics* (CSM1 169).

For example, light reflected from an apple hits the eye and causes motions inside the eye. These motions cause further motions in the nerves that propagate toward the brain. In the brain, the mind intervenes, and a first form of awareness occurs: the feeling of the body being affected in a certain way—what would now be called "sensation" (as will be detailed in chapter 3). After this second step, in a third stage, a form of awareness will come about that concerns the external causes of the sensations. The perceiver would now be aware of the apple they are looking at. There is a process from external cause to awareness of bodily feeling caused by it, to awareness of external cause.

2.3.5 The Union of Mind and Body

Descartes's views were scrutinized and criticized by his contemporaries and have been so ever since. Descartes clearly thought mind and matter were profoundly different, but he also held that, in a conscious human being, mind and body were united. In fact, Descartes stresses the union between mind and body runs deep. Together, they form a single person who feels, sees, imagines, and thinks. Descartes expresses this in a famous passage in his *Sixth Meditation*:

> Nature also teaches me, by these sensations of pain, hunger, thirst and so on, that I am not merely present in my body as a sailor is present in a ship, but that I am very closely joined and, as it were, intermingled with it, so that I and the body form a unit. If this were not so, I, who am nothing but a thinking thing, would not feel pain when the body was hurt, but would perceive the damage purely by the intellect, just as a sailor perceives by sight if anything in his ship is broken. Similarly, when the body needed food or drink, I should have an explicit understanding of the fact, instead of having confused sensations of hunger and thirst.

> For these sensations of hunger, thirst, pain and so on are nothing but confused modes of thinking which arise from the union and, as it were, intermingling of the mind with the body. (AT VII, 81; CSM2, 56)

But how does this union come about? If pure thinking is fundamentally different from matter, yet can also be coupled to it—as in conscious perception—how does such coupling happen? Given Descartes's insistence on the distinctness of the two substances—extended and thinking—the question of how such interaction is even possible arises. Notoriously, Descartes pointed to the pineal gland in the brain as playing a special role in the union between mind and brain. One reason for the choice for the pineal gland was that Descartes pictured it as moving in step with both sensory stimulation and imagination. Another was that it is one of the few cerebral structures that doesn't exist in a left-brain and right-brain version (Lokhorst, 2013).

As we have one consciousness, and not two, the choice for the pineal gland might have been appropriate. But pointing to a structure as playing a crucial role in connecting mind and body doesn't address the fundamental question of how the intermingling of essentially different substances is possible in the first place. A prominent critique of Descartes's mind/body position has become known as *the interaction problem.*

A very famous early expression of the interaction problem was given by Princess Elizabeth of Bohemia (1596–1662), in an exchange of letters with Descartes. The principal problem that Elizabeth raised was how it was possible that something that is not extended in space, namely the mind, was able to interact with something that *is* extended in space. Elizabeth knew that Descartes claimed that the pineal gland was the locus where the mind set in motion the body, via its nerves. But while the transfer of motion can be perfectly understood as resulting from the impact of one physical object upon another one, such understanding seems to vanish once the impact comes from something not extended in space.

Descartes responded with the claim that a special mode of causation applied, one that should not be modeled on causation in the physical realm (Wilson, 1978, p. 213). Those who raise the interaction problem, so Descartes states, make the mistake of supposing that the way the mind affects the brain and the brain the mind must be like mechanical causation. But mechanical causation happens between extended, material things and should not be taken as applying to the way the mind forms a union with the body, so Descartes held. The interaction problem relies on a wrong picture.

But what, then, is the right picture? Descartes never provided that in any detail. This might have been due to confidence in his own method. For if Descartes was convinced that he had proved with absolute certainty that mind and body were different, the question of how they "intermingle" would only be of minor importance to him.

Descartes's dualism rests thus on his system as a whole, which includes his method and the justification for it. Without the powers to come to certain conclusions based on a priori reasoning, Descartes's whole construction becomes wobbly. In her correspondence with Descartes, Elizabeth also tracks this vulnerable spot in Descartes's dualism. She questions the certainty that Descartes expresses concerning his own reasoning. How can Descartes be so confident that his thinking and conceiving is right? How can he be so certain that the absolute separation between thinking and matter he thinks he can make in thinking reflects a real distinction? Elizabeth points out that what seems evident to Descartes is not so to her. She writes:

> I admit that it would be easier for me to concede matter and extension to the soul than to concede the capacity to move a body and to be moved by it to an immaterial thing. (AT III 685/S 68, Shapiro, 2007, p. 68)

The broader point Elizabeth makes concerns the reach and limits of a priori thinking. Descartes held that people have a God-given capacity for rational thinking. Exercising it would lead to absolute truth in mathematics and in reasoning about reality. Human reason is seen here as absolute, fixed and unchangeable. But if it was so, why would Elizabeth, by using her reason, find it possible to conceive what Descartes declares impossible to conceive, namely, that matter can think? Elizabeth's concern can be amplified by noticing that what human thinkers find conceivable varies in historical periods. Many ancient and medieval dualistic philosophers would have agreed with Descartes that it is inconceivable that machines think, but in the twenty-first century, the idea of artificial intelligence (AI) has found a firm footing. Many people might still deny that an AI is really intelligent, but not necessarily because they are convinced that the idea of material intelligence is unthinkable. Cartesian apriorism doesn't even seem to hold for mathematics when this is seen in a historical perspective. Time and again, what previous mathematicians declared to be inconceivable was conceived and developed: negative numbers, the square root of negative numbers, and different forms of infinity, to mention only some examples.

This leads to another question that can be raised with respect to Descartes's system. If the whole system is based on a particularly pure form of ideas—the one he finds in mathematics—where do these ideas come from? Many thinkers have been unhappy with leaving this question unanswered. One of them is the pragmatist philosopher John Dewey (1859–1952). He wrote: "'Reason' is not an antecedent force which serves as a panacea. It is a laborious achievement of habit needing to be continually worked over" (Dewey, 1922, p. 198).

This passage can be read as a call to explain where ideas come from. It refuses to see reason as something absolute that has always been there, and asks to find out how thinking came from nonthinking. It is a question for the origins of mind. We'll see this question reappear again and again in subsequent chapters.

Bibliographical Note for Chapter 2

Descartes scholars complain that the standard treatment of Descartes in introductions to the philosophy of mind is often focused on "the argument from doubt" (Rozemond, 1998, p. 19; Wilson, 1978, p. 190). The argument from doubt starts from two premises: I can doubt that the material world, including my body, exists. I cannot doubt that I am thinking now. Then it adds that identical things must have all properties in common—this is called "Leibniz's law." But, so the reasoning goes on, if my body's existence can be doubted, but not the existence of the thinking "I," the thinking "I" cannot be identical to the body. Then it is shown how the argument depends on a logical mistake, roughly because "being doubtable" or "not being doubtable" is not a property of the things being doubted, but of the doubter. So, with respect to "doubtable" and "not doubtable," Leibniz's law does not apply. As can be gleaned from the current chapter, the motivations for Descartes's dualism are richer. Equally, many Descartes scholars regret that the epithet "Cartesian" has come to stand for much that allegedly has gone wrong in relatively recent thinking about the mind. They will emphasize, for example, that Descartes was very much concerned with how mind and body formed a union (see, e.g., Hatfield, 2019, pp. 106–107).

There are contentious issues of interpretation for all philosophers treated in this chapter. Readers might be puzzled by how Aristotle could ascribe perception to animals, while he also says that perception has abstract aspects and therefore requires the rational soul, which is only to be found in humans. This puzzle can be solved if Aristotle thought of perception as possible both without and with abstraction. Animals would then have a kind of nonconceptual perception, while in humans, some form of conceptualization takes place as well (see Bäck, 2008). A much-discussed related issue is whether Descartes really denies all forms of sensation to animals. Sometimes he seems to endorse that, whereas in other places, he asserts that animals enjoy some kind of sensation. So, does he allow for a sort of sensation that is mental or conscious without the involvement of the *res cogitans*? Opinions are divided, as can be learned from Harrisson (1992).

Throughout his *From Soul to Mind. The Emergence of Psychology from Erasmus Darwin to William James*, Edward Reed stresses that in many European nations, until as late as roughly the twentieth century, publicly renouncing dualism or endorsing materialism was dangerous, illegal, and punishable (Reed, 1997). Despite that, there were some notorious materialists. Foremost of these is the British philosopher Thomas Hobbes, who was cited in this chapter for his opposition to Aristotelianism. Hobbes came under investigation at some point but escaped punishment. He argued that materialism is compatible with Christian theology. The eternal life would happen because of the resurrection of physical bodies. The prolific Margaret Cavendish (1623–1673), Duchess of Newcastle-upon-Tyne, was another notorious materialist, sharing Hobbes's views that mental phenomena like perceiving and thinking are processes in the brain. (Cobb, 2020, pp. 41–42). Like the well-known French materialist Julien Offray de La Mettrie (1709–1751), she believed in the existence of some fundamental vital force in nature—a special force that animates living things—so her materialism was different from the mechanistic views of Hobbes.

Overall, some form of dualism remained dominant in Western culture until well into the nineteenth century (Reed, 1997). The precariousness of defending materialist views can be seen by considering the reactions to a passage in John Locke's otherwise celebrated *Essay Concerning Human Understanding*, in which he writes:

> We have the ideas of matter and thinking, but possibly shall never be able to know whether any mere material being thinks or no: it being impossible for us, by the contemplation of our own ideas, without revelation, to discover where Omnipotency has not given to some system of matter, fitly disposed, a power to perceive and think, or else joined and fixed to matter, so disposed, a thinking immaterial substance: it being, in respect of our notions, not much more remote from our

> comprehension to conceive that God can, if he pleases, superadd to matter a faculty of thinking, than that he should superadd to it another substance with a faculty of thinking. (Locke, 1690/1975, Book IV, chapter 3, § 6, pp. 540–541)

Locke was vehemently attacked for this passage and accused of being a materialist and atheist (Cobb, 2020, p. 44). This is surprising, given that Locke does not in any sense defend materialism here. He just states—like Elizabeth of Bohemen had done before in a sharper way—that dualism cannot be proven a priori.

Though thought and rationality remained main motivators for dualism, there were references to sensations too as creating particular problems for materialism in the seventeenth, eighteenth, and nineteenth centuries (for some examples, see Cobb, 2020, chapter 2, pp. 43–44).

3 Inside the Mind: The Emergence of Scientific Psychology

> In modern neuroscience, our working definition of perception is captured well by the Oxford English Dictionary: "The action of the mind by which it refers its sensations to an external object as their cause." This definition has roots in the corpus of eighteenth-century philosophy—beginning with George Berkeley and David Hume—was expanded upon by later British associationists, and became a foundation of both experimental psychology and modern neuroscience.
>
> —Thomas Albright (2015, p. 22)

3.1 From Mind to Psychology

As is clear from the previous chapter, thinking about the mind has a long history. Many diverging—and sometimes converging—ways of reflecting on the mental can be discerned throughout the history of civilization, even if one only considers Western civilization. Moreover, even in the same historical period, thinking about the mind took many forms. Since the eighteenth century, people have started to use the word "psychology" to designate ways of thinking about the mind. Originally being used in contexts such as involving "knowledge about spiritual creatures" (Danziger, 1997, p. 37), the word eventually became associated with an academic discipline aimed at investigating the mind in a *scientific* way. That is, in the nineteenth century, psychology became an institutionalized academic discipline—complete with curricula, degrees, professorships, laboratories, and funded research programs.

Scientific academic psychology was not a firstborn. When it saw the light of day, other scientific disciplines like physics, chemistry, and physiology were already well established. Indeed, the idea of a scientific psychology and its practice, from early on, has often been inspired by analogies to those sciences. This inspiration can go from sharing an ideal of being objective, to conceiving of the mind in terms of mental objects that combine and interact in lawful ways, to be discovered by the science of psychology though observation and experiment.

Thinking about the mind in terms of interacting with mental objects or inner entities has many and old roots. In Western culture, one can see it was clearly present in the writing of some ancient Greek thinkers. But in the period of thinking called "modern," it became dominant. As we have seen in the previous chapter, Descartes was committed to it. It was further endorsed and elaborated by later thinkers, such as in the massively influential writings of John Locke (1632–1704) and David Hume (1711–1776). Besides explicitly picturing the mind in terms of "mental atoms" and combinations thereof, both thinkers also expressed allegiance to two further views with a long pedigree in thinking about the mind: *mediationism* and *representationalism.* According to *mediationism*, both perception and thought, though they seem to be directly about the world, are only indirectly so. It might seem to us that when we see an apple, we see the real apple directly, and perhaps also that when we think about the moon, it is the heavenly object itself that we relate to. Mediationists hold that such opinions are naïve and mistaken. In perception, what stands before the mind are not the things seen but seeings, not things thought about but thoughts. Mediationism can be coupled to *representationalism.* Such happens when it is held that the mediating seeings and thoughts are representations of objects instead of the objects seen.

It is one thing to theorize about the mind in terms of mental objects; it is another one to investigate these objects and their ways of behaving scientifically. Since the scientific revolution, this meant: experimentally, observationally, and mathematically. The sciences of physics, chemistry, and anatomy had spectacularly progressed since the scientific revolution, this raised hopes for the science of psychology. But it took broader developments, starting in physiology, to see the emergence of what became known as "experimental psychology."

Though experimental psychology added a lot of scientific content to ideas from thinkers like Descartes, Locke, and Hume, it still kept their basic way of thinking about the mind. That is, standard experimental psychology remained committed to the mind as containing mental objects. Some mental entities are simple (sensations), others are complex (perceptions). The more complex are derived from the simpler. Mental entities mediate between the world and the organism by representing the world. Indeed, this picture of the mind, or something close to it, is still widely shared in the current sciences of the mind—from perceptual psychology to cognitive neuroscience.

Despite its lasting influence, this framework for thinking about the mind has been the object of varied and repeated criticism. It has been attacked in its various aspects and in its totality. For example, it has been argued that mediationism implies that contact with the real world has irretrievably been lost, or that there is no evidence whatsoever, only mistaken theoretical motivations, for the existence of mental atoms. These criticisms had early proponents, such as Tomas Reid (1710–1796), and have continued to resurface in thinking about the mind ever since.

3.2 The Empiricist Picture of the Mind

As seen in the previous chapter, Descartes made a complicated attempt to draw mechanistic thinking into domains that had often been regarded as mental—such as animal feelings or human sensations. Indeed, though Descartes kept a separate space for uniquely human reason, he went quite a way in mechanizing the mind. He accounted for perception–action links in mechanistic terms. According to Descartes, such fully mechanical action could occur in humans too, even if the mind can intervene. The full human mind operates via the intervention of the intellect in what remain automatic processes in animals. Conscious sensation and conscious perception are a result of this combination of bodily and intellectual processes. Perception is a process in which the mind first becomes aware of how the body is affected via external causes, as when light hits the eye, or via internal causes, as in a stomachache. On the basis of bodily feeling, humans become aware of the causes of those bodily feelings. Unlike the sensing of bodily feeling, perception is about the world.

Descartes was not the only one who thought about perception of the outside world as derived from body-based sensing. Anatomists and physiologists were investigating the body and brain within a similar framework, though with an additional focus on specifics of physiology and anatomy. For example, Thomas Willis (1621–1675), a prominent physician from Oxford, engaged in the dissection of a wide range of animals, from fish to monkey. He thought of perception in terms of a sensible impression that was formed in the eye and then traveled to the brain where perception would arise, while the perceptual image would be stored for memory elsewhere in the brain (Cobb, 2020, p. 39).

The British empiricists, with as main representatives John Locke (1632–1704), George Berkeley (1685–1753), and David Hume (1711–1776), in no sense engaged in anatomy or physiology. They were primarily concerned with knowledge (Richards, 1992). In particular, they addressed the question of what humans can and cannot know and have knowledge about. Famously, they were *empiricists*, claiming that, in the end, all knowledge derives from what is apprehended by the senses. They opposed Descartes's view that clear and distinct ideas formed a separate and superior source of knowledge. In exploring the theory of knowledge, or engaging in epistemology, they painted a picture of the mind which would have lasting influence.

The empiricists conceived of the mind in terms of mental atoms and compounds thereof. They distinguished between simpler and more complex mental entities. Basic mental entities arise from stimulation of the sensory organs or, added Hume, from the inside, as when we feel pain and pleasure. From combinations or operations on basic mental entities more complex mental entities derive. So, according to Locke, the mind forms the more complex idea of a cat by combining colors and shapes, or the general idea of a cat by subtracting all those qualities that are not shared by all cats. They also hold that the mind is able to consider several ideas together, so that it can assess relations between ideas ("this cat is bigger than that cat" or "cats are bigger than mice").

Like Locke, Hume admired the physics of Isaac Newton. Hume thought that, similar to how worldly "bodies"—the contemporary term for "objects"—interacted according to physical laws, items in the mind influenced each other in regular ways. These mental objects he called perceptions. Perhaps contrary to our expectations, the name perception covers, for Hume, all that is to be found in the mind. It includes the visual perception of a tree, a pain in the foot, or a thought about the weather. At the root of all perceptions stand what Hume calls "impressions of sensation." These arise from the five senses, but as said, also include pain and pleasure. Ideas are understood as "faint copies" of impressions that occur in, for example, thinking or memory. The pain you are having now is an impression, but if it subsides and you later remember it, you will deal with an idea of the pain.

Besides impressions of sensation, or "original impressions," Hume posits "secondary impressions." They arise when the mind reacts to ideas, as when your memory of a pain instills a fear. Importantly, Hume thought there was a class of simple impressions, which could then be combined into more complex impressions and lead to complex ideas. Color was thought of as a simple impression because it allegedly could not be further analyzed into component impressions. The impression one has when observing a red rose, or the idea of a golden mountain when imagining, can both be seen as composites of simple impressions or simple ideas of color and shape.

Besides mental counterparts for the bodies of the natural sciences, Hume described mental analogues for laws of nature: principles of laws of association, present in how impressions and ideas interact. A first principle of association is that perceptions, in the Humean sense, tend to evoke perceptions that resemble them. So, if I see a lute, I might think of a guitar. Contiguity in space and time form associative principles too: I might think of Mexico when I think of California, or of summer if I think of spring. The most useful principle of association, however, according to Hume, is the one that connects causes and effects, as when one thinks of eating spoiled eggs and getting sick. Hume sees this principle at work in the most advantageous forms of human thinking, because it forms the basis of sound empirical scientific reasoning.

Locke expressed the thought of the entities inside the mind as representations that mediated our access to the world, both in perception and in thinking as follows: "It is evident the mind knows not things immediately, but only by the intervention of the ideas it has of them. Our knowledge, therefore, is real only so far as there is a conformity between our ideas and the reality of things" (Locke, 1690/1975 Bk IV, Ch. 4, § 3, p. 563).

The mediationism Locke here expresses implies that when we see an object, we are not in direct visual contact with the object, but only via an intermediary mental entity—an idea of the object. And the same goes for thinking: Our thought about objects does not directly reach out to the objects but only indirectly via a thought. If the mind eventually arrives at the object of perception or thinking, this is because the inner entity—the idea in perceptual experience or in thinking—conforms to or represents the object.

As he states in the quote just presented, Locke finds this mediationism self-evident. As Nichols and Yaffe (2023) note, this is typical for endorsements of mediationism in broadly Locke's times: Mediationist claims are presented as obvious truths, not standing in need of explicit argumentative support. Possibly Locke has in mind that mediationism must be true, because, when I see a cat, the cat itself is not in me or in my experience. Therefore, what's in me or in my experience must be something different than the cat, namely a representation or semblance of the cat. Indeed, in what Nichols and Yaffe consider to be one of the few examples of an argument for mediationism, David Hume seems to be playing that card. First, Hume makes the following claim: "Nothing can ever be present to the mind but an image or perception" (Hume, 1748/1999, section 12, part 1, § 9, p. 114). Next, Hume provides the following argument: "The table, which we see, seems to diminish, as we remove farther from it: but the real table, which exists independent of us, suffers no alteration: it was, therefore, nothing but its image, which was present to the mind. These are the obvious dictates of reason" (Hume, 1748/1999, section 12, part 1, § 9, p. 114).

Hume here clearly follows the thought that it can't be the table itself you have in your mind when you perceive it. For the table itself does not get smaller when you move away from it. Then he draws the conclusion that it must be something else that is in your mind. That something else, so Hume says, must be an image. Interestingly, like Locke did, Hume expresses his absolute confidence in his reasoning and its conclusion by writing that it follows "the obvious dictates of reason."

But despite Hume's confidence, one can question his reasoning. Is it indeed impossible that one sees an objective table instead of an inner image because the table looks smaller when one walks away from it? Should one have to resort to a mediating image to explain the changing looks of the table? We'll shortly see that critics of mediationism claim there is a better alternative to Hume's reasoning and conclusion. Of course, there are additional reasons mediationism might seem obvious to thinkers such as Locke and Hume. The invocation of mental images to account for experiences of absent items has a long pedigree in the history of thinking. To give just one example, Aristotle thought of the imagination as the capacity to produce mental images (Shields, 2020, Supplement on "Imagination").

Just like images have been invoked to account for imagination, they can be invoked to explain perceptual error. If it seems to one that one is seeing a cat, while there is only a shadow, isn't the best explanation that one has a visual image of a cat, caused by a shadow?

In Locke's words, one can say that in adequate perception the image or idea of the cat "conforms to the reality of things," while in the case of perceptual error it doesn't (see the quote from Locke a few paragraphs back). As noticed, Locke has more to say about what it means for an idea to "conform to the reality of things." In writing this, he gave a canonical name to the distinction between properties of an object of experience that are in objective reality as they are in experience—these Locke calls "primary qualities"—and properties of an object whose appearance in experience is different from how they are in objective reality—these Locke names "secondary qualities." As mentioned, the distinction was thematized by Galileo, Descartes, and Newton, among others. Examples of primary qualities are shape, motion, or quantity. Color and taste are examples of secondary qualities.

The conception of color as a secondary quality gives rise to a picture of the perception of color that is very different from Aristotle's view on the same matter. In Aristotle's view, perceiving the color of an object is being informed by the same form that inhabits the object. Something objective is shared between the object and the perceiver. In a secondary qualities account, there is a causal chain beginning at the object and ending at the perceiver, but besides the causal connection, nothing is shared between the first and the last link in the chain.

The secondary quality account can be accompanied by a view of color experiences and experiences of other secondary qualities as arbitrary. Since these experiences don't resemble their causes, they could be anything—as long as they retain the right causal, correlative, relation to their objective causes. Locke develops the contingency of experiences of secondary qualities with what has proved to be one of the most impactful thought experiments in recent philosophy, which has become known under the name of "spectrum inversion." It concerns the possibility that, in the words of Locke, "the same Object should produce in several Men's Minds different Ideas at the same time; v.g. the Idea, that a Violet produces in one Man's Mind by his Eyes, were the same that a Marigold produced in another Man's, and vice versa"'(Locke, 1690, Bk.II, Ch. 32, § 15, p. 389).

With Locke and Hume, we find two philosophers who refine a picture of the mind painted in terms of internal private mental objects, subject to laws, regularities, or rules. These philosophers contribute to a process of objectification, internalization, and privatization of what's inside the mind.

As said, the British empiricists were mainly interested in the theory of knowledge, what is now called epistemology. They weren't setting up experiments or theorizing how impressions related to the brain (Richards, 1992). However, the picture that animated the philosophies of the British empiricists, of the mind as containing mental elements and operations, would become adopted by the emerging new science of psychology.

Although various strands of modern thought laid the groundwork for a science of the mind—envisioned as comparable to natural sciences like physics and chemistry—not all philosophers embraced this perspective. One notable exception can be found in the work of Immanuel Kant (1724–1804). Kant argued that the study of the mind could *not* go the way of physics. The first of two prominent reasons he gave concerned the subjectivity of experience (see the treatments in Gardner, 1985, chapter 4; Käufer & Chemero, 2021, chapter 2; Reed, 1997, chapter 2 and throughout). In Kant's view, experience is tied to its experiencer in ways in which external objects aren't. Two persons can experience the same tree but never experience the same experience. Second, experience is experienced only in time. The objects of physics, are, besides in the single dimension of time, also in the three dimensions of space and can therefore be studied with mathematics. But experience's one-dimensionality defies such mathematization.

Kant also saw an obstacle to a scientific study of the mind in the special ways in which living organisms exhibit goal-directedness in their structure and behavior (Kant, 1781/1999 and 1790/2001; see discussion in Withagen, 2022, pp. 17–18). In his view, organisms are essentially different from mechanical machines. He saw science as necessarily mechanistic and concluded that science could not account fully for the purposeful activities of organisms. Kant was aware that machines show purpose in their structure, but argued there was an essential difference between the functional dependence that exists between the parts of machines and the teleological makeup of organisms. One part of a watch sets another part of the watch in motion, but its existence isn't caused by the other part. But in organisms, parts depend on each other for their existence: One part exists because of the other part, and vice versa. The heart exists because it gets oxygen from the lungs, and the lungs exist because they get blood from the heart. The interdependence between parts of organisms concerns their continued existence. The parts of a living organism cause the whole to continue to exist. As we'll see in chapter 11, Kant's ideas have become revived in enactivism (Weber & Varela, 2002).

Kant later argued that, instead of being modeled on physics, a science of the mind should be historical and anthropological—based on descriptions of concrete contextualized activities of different kinds of individuals and groups (Kant, 1798/2006). Such anthropology would be practical in that it would provide insights relevant to how we should act toward each other. Kant's idea of psychology-as-anthropology therefore sees it as inherently practical and as having normative or ethical significance.

3.3 From Physiological to Experimental Psychology and Beyond

3.3.1 Bringing Sensation into Science

Despite Kant's doubts, psychology, as an academic scientific discipline, modeled on the natural sciences, became a reality. It should not be forgotten that, while the philosophers discussed so far were active, medical writers and physiologists were making vital contributions—through empirical investigations of the body and brain, as well as through theoretical explorations. A constant stream of physiological thinking—tapped into by Descartes—was developed (Cobb, 2020; Finger, 1994; Richards, 1992), by people like the celebrated anatomist Andreas Vesalius (1514–1564), the already mentioned Thomas Willis, or the microscopist Jan Swammerdam (1637–1680). Erasmus Darwin (1731–1802), grandfather of Charles Darwin, devised a theory of human behavior that synthesized associationist empiricism with contemporary physiological knowledge—a science that focused on movements and on how they were caused by what was designated with the newly minted term "stimuli," that could either arise externally or internally (see Danziger, 1997, p. 55; also Richards, 1987).

Erasmus Darwin's fusion of the science "of stimulated motion with philosophical empiricism had considerable potential, but Darwin was ahead of his time" and "forgotten by his countrymen," Danziger notes (Danziger, 1997, p. 57). This injustice was restored to some degree by the attention Erasmus Darwin received in the very influential book by Johannes Müller (1801–1858) titled *Elements of Physiology* (Müller, 1833–1840/1837–1843). Müller's text dealt extensively with sensation and perception in the context of the nervous system. For example, Müller speculated that the specific ways in which experiences in the different sensory modalities are experienced—as auditory versus visual, for example, was due to "specific nerve energies" (Müller, 1833–1840/1837–1843

Book V). In Britain, several texts appeared under the banner of "mental physiology" (Danziger, 1997, 52).

Shortly after the publication of his *Elements of Physiological Psychology* (1874/1904) Wilhelm Wundt (1832–1920) opened his Institute for Experimental Psychology in Leipzig, Germany. It has become customary to take this as the birthdate of psychology as a self-standing scientific discipline. Wundt institutionalized a tradition of foundational theorists of psychology, often trained as physician or physicist. They include Ernst Weber (1795–1878), Gustav Fechner (1801–1887), and Herman von Helmholtz (1821–1894). They investigated perception by investigating sensation. These theorists studied sensation in a *psychophysical* way, namely by relating how for example sensations of color or tone changed with precisely controlled changes in the physical properties of the stimulation delivered to the sensory organs. Early experimental psychology had this focus on sensory psychophysics, so Lawrence E. Marks states, "in large measure because of the widely held view, empiricism, that the contents of the mind derive from sensations" (Marks, 2011, p. 29).

By careful investigation of how increases in the intensity of the physical dimensions of a stimulus—for example, a light that is made to shine stronger—led to changes in sensations—of brightness, in our example—these early psychophysical psychologists arrived at fundamental psychophysical laws now known as "Weber's law" and "Fechner's law." These two fundamental laws describe the relation between physical stimulus and sensation. For higher levels of physical stimulus intensity, one needs to add increasingly more intensity for a change to be reflected in sensation. A small increase in sound energy is easily perceived, but for a more intense sound stimulus, a significantly greater increase in energy is required to produce a notable change in sensation.

With the work of Edward Titchener (1867–1927), a British-American disciple of Wundt, the psychophysical program became self-consciously *structuralist*. Reminiscent of the Humean or empiricist distinction between simple and complex perceptions or impressions, structuralism aimed at finding the atomic building blocks of sensory consciousness. Moreover, the project was explicitly conceived of as analogous to physics and chemistry. Operating under this banner, structuralists analyzed several types of conscious experiences—sometimes with the ultimate aim of providing the complete catalogue.

In contrast to what Hume had held, psychophysicists showed that colors are not experiential atoms but could be analyzed themselves as containing an internal structure. That is, specific sensations can differ along the dimensions of hue, saturation, and brightness. Hue concerns what we ordinarily speak about as "colors" that things have in the sense that an object can be red, blue, green, yellow, or orange. But a red object can have more or less of the same red in it; it can vary along the dimension of saturation. Moreover, a light of the same color can be made to shine stronger or be dimmed. Then you'll experience the same hue and saturation combination as brighter.

In fact, psychophysicists aimed at identifying *all* the ways or dimensions in which conscious perceptual experiences could differ—striving for a level of completeness matching that provided by the periodic table in chemistry. In this vein, they discerned very general dimensions such as quality (e.g., color), form, or duration. Psychophysicists discussed which dimensions exactly allowed to completely analyze sensory experience and such matters as whether properties such as pleasantness formed a dimension characteristic of sensations themselves, or rather belonged to feelings that accompanied sensations (Marks, 2011, p. 29).

But structuralist psychophysics allows for more than a dissection of individual sensory experiences. It also makes it possible to systematically investigate and represent the *relations* between classes of sensations. Such can be done for color by drawing a color solid. This is a three-dimensional figure on which ideally all possible color sensations are charted in such a way that there is a systematic correspondence between the distance of the location of a color on the figure and how much the experiences of those colors are felt to be similar or dissimilar. The color solid has been invoked by later philosophers as rendering the structure of color experience or sensory consciousness and thereby explaining consciousness (Clark, 1993).

Titchener, the father of structuralism, made it clear that, in his view, the "psycho" in psychophysics was there for a reason. He stressed that the object of psychophysics were properties of sensations, which he saw as quite different from properties of stimuli. Such thinking took for granted the distinction between physical properties and what with Locke were termed "secondary qualities." Experienced shape is like real shape. But color is a secondary quality, because nothing in the wavelength of light that we see as red is itself red. Or, to repeat what Newton wrote: "The Rays properly speaking are not coloured" (Newton, 1730/1952, Bk. I, Pt. II, Prop. II, Thm. II, Definition,

p. 124). Nothing in the world corresponds, or is similar to, the red we find in experience. Experience of secondary qualities is, as far as its felt character is concerned, *sui generis*. Titchener gave further expression to such a view by warning for what he called "the stimulus error":

> We are constantly confusing sensations with their stimuli, with their objects, with their meanings. Or rather—since the sensation of psychology has no object or meaning—we are constantly confusing logical abstraction with psychological analysis; we abstract a certain aspect of an object or meaning, and then treat this aspect as if it were a simple mental process, an element in the mental representation of the object or meaning. . . . We do not say, in ordinary conversation, that this visual sensation is lighter than that, but that this pair of gloves or this kind of grey paper is lighter than this other. We do not say that this complex of cutaneous or organic sensations is more intensive than that, but that this box or package is heavier than this other. We do not even say, as a rule, that this tonal quality is lower than that, but rather that this instrument is flat and must be tuned up to this other. Always in what we say there is a reference to the objects, to the meaning of the conscious complex. It is not the grey, pressure, tone, that we are thinking of; but the grey of leather or paper, the pressure of the box, the pitch of the violin. . . . What is more natural than to read the character of the stimuli, of the objects, into the "sensations" with which certain aspects of the stimulus or object are correlated? (Titchener, 1905, pp. xxvi–xxvii)

Titchener is characterizing sensations here explicitly as meaningless, and the way they feel as unconnected to the things we see in the world. He thereby expressed a view widely shared by mechanistic thinkers and the founders of experimental psychology.

3.3.2 The Next Stage: Perception

Much of early experimental psychology was focused on sensations. But, of course, sensations can't have the last word when it comes to studying perceptual experience. It is animals, persons, smiles, chairs, rooms, or soccer games that we perceive. Naturally, psychology could not avoid detailing how, over and above sensation, perception happens.

We have seen that, in his account of perception, Descartes held that external stimuli, like light, first gave rise to events inside the organism. These events, to which Descartes sometimes referred as "brain motion" (Withagen, 2022, p. 13), only concerned bodily conditions. The same scheme was later widely accepted and couched in terms of a relation between sensation and perception. External objects are perceived via the prior receiving of sensations. In fact, often this picture contains the idea that there is both a causal and an epistemological order—one concerning knowledge—in the sensation-perception relation. Such order can be epistemological, sensations being the evidential basis for perception. On the causal reading, sensations are seen as a first stage in a temporal and causal process of which perception forms a later stage. On the epistemological reading, we have more certainty, or better knowledge, about sensations than we have about objects. The epistemological and temporal orders go together naturally. This causal and epistemological view of sensations is present in Hume's empiricist atomism, and arguably in Locke's views too (see Hatfield, 2021 on Hume; Uzgalis, 2022, section 2.2, for some nuance about Locke).

In any case, one can see both the causal and epistemological version at play in the work of John Stuart Mill (1806–1873). Mill, influenced by William Hamilton (1805–1865), endorsed that what we experience primarily and with certainty are sensations, such as patches of light and color. He expresses this view as follows: "The sense of sight informs us of nothing originally, except light and colors, and a certain arrangement of light and colors" (Mill, 1842/1864, p. 165).

But then the mind operates on these sensations to form perceptions that concern their external causes. Mill proposed to think of these mental operations as a kind of reasoning. More specifically, he argued that the link between sensation and perception was brought about by inferences. Perceptions are the result of an inferential process that takes sensations as its premises and delivers perception as its conclusion. For example, from certain sensations of light and color a mental inferential process leads to the perception of a falling star.

That is, given that certain sensations occur, the mind posits certain events in the outside world as causing precisely these sensations. Think mystery novels. Certain traces are left, and, considering all these together, the heroine of the story concludes what must have happened for exactly these traces to be present. Mill thought that our minds work in the same way when our sensory organs are stimulated: Sensory nerves are activated, leading to sensations, and on that basis our minds posit that such and such must be present in the outside world to cause that particular pattern of sensations. While the detective in the story might reason consciously, the mental operations invoked by Mill are taken to occur below the radar, or unconsciously. They are *unconscious inferences*.

The notion of perception as unconscious inference became massively influential in scientific psychology, thanks to the work of the physicist and psychologist Hermann von Helmholtz (1821–1894). Helmholtz, having studied John Stuart Mill's writings on the topic, strongly endorsed that sensation constitutes a first stage—one that causes the subsequent stage of perception (see Marks, 2011; Reed, 1997, chapter 7).

Helmholtz took a traditional starting point: Sensations and perceptions were effects within the mind of causes outside the mind. He also explicitly claimed that, in perception, perceivers experience objects *as* the causes of their experiences. When you perceive your laptop's screen in front of you, you experience it as an entity in the world beyond your skin, more specifically, as an objective screen that is causing the experience of a screen in you. Helmholtz clearly believed that perceptual experience is about the extradermal, external world. Yet, he considered this power to reach beyond the brain, or beyond the skin, as something remarkable. For he reasoned that it must be based on what is in the brain only: sensations. It is through the application of mental operations, in particular, through unconscious inferences, that it is possible to bridge the gap between sensation and outer world.

Again agreeing with Mill, Helmholtz took the mental operations linking sensation to perception to occur unconsciously. As in Mill, they are unconscious inferences. According to Helmholtz, such unconscious inferences rely on knowledge acquired during previous experiences. For example, buildings cast shadows. People will have experienced this many times in the past. When we only see the shadow, we might infer based on this past experience that there's a building causing his shadow. Something like that goes on unconsciously when we receive sensations. The sensations we have are effects of objective causes. A tree causes sensations. In seeing, we unconsciously infer from the effect (sensations of a tree) to causes (the tree out there). The mental operations of inference that help to bridge the gap between sensation and perception rely on knowledge gained through experience. What is experienced as the result of this process corresponds to what, given the perceiver's experience, is the most likely cause of her sensations (Patton, 2018).

3.3.3 Beyond Perception: Memory and More

Locke, Hume, and the tradition that followed them were interested not only in perception but in all cognitive and mental phenomena, including memory, imagination, and thought. Locke can serve as an example here. He drew a distinction between ideas of sensation and ideas of reflection.

Ideas of sensation were received through the senses. Ideas of reflection arise when the mind operates on ideas of sensation. Such mental operations can be varied. The mind can "revive" previously had ideas—this forms the basis of memory (Locke, 1690/1975, chapter II, book viii, section 4). The mind can combine simple ideas into complex ideas. It can juxtapose ideas and form a new idea of the relation between the compared ideas. Abstract ideas are formed by retaining from individual ideas only common features (see Locke, 1690/1975, chapter II, book 10, section 2). In these ways, Locke accounts for cognition and mentality in terms of a restricted set of operations performed on representations. We have seen in this chapter how such a view has been espoused and elaborated by theorists like John Stuart Mill and Hermann von Helmholtz. Understanding the mind in terms of mental operations on mental representations also forms the core of computational views of the mind, that are influential in many domains, including neuroscience, psychology, and artificial intelligence (see chapter 7). These approaches see not only perception but also mental faculties such as memory and thinking in representationalist terms. Representation is the common coin of the mental. Not only seeing but also remembering or thinking is to represent. The common coin allows for intellectual trade among perception, memory, and thought, so that what we see can influence what we remember and think and vice versa—as in Helmholtz's account of unconscious inference.

Nichols and Yaffe describe the continuity at issue as a " set of commitments . . . drawn from the work of Locke, Berkeley, Hume and many others going back to Plato and Aristotle. Its principle commitment is to mental representations, called 'ideas' or 'images', that are believed to mediate all our experience of the world—from conception to memory and perception" (Nichols & Yaffe, 2023, section 2).

Nichols and Yaffe (2023, section 3.1) point to Rorty (1979) to make the further claim that mediationist representationalism is "the key player in the story of Early Modern Western philosophy . . . which would continue as the story of contemporary analytic philosophy."

But not everybody joined this tradition.

3.4 Not Impressed

The style and content of thinking about mind and perception, as in the tradition from Descartes through Helmholtz, remain very much alive and forms the core of contemporary mainstream approaches both in philosophy and in the sciences. Yet all of the traditional tenets have been criticized from its inception. Many of its claims have come under scrutiny and still are contested. In the remainder of this chapter, we will focus on criticisms of mediationism, of sensation as understood in philosophy and psychophysics, and of unconscious inference. The focus will be on criticism formulated before or at the beginnings of the twentieth century. In future chapters more recent and contemporary critics will be dealt with.

3.4.1 Mediationism and Representationalism

If our access to the external world is always mediated by internal images, impressions, ideas, or representations, why assume that we do access the world outside the mind at all? This is a fundamental and returning question critics have addressed to mediationism.

I see an apple. According to mediationism, this happens because I have certain visual sensations and a perceptual visual representation built from these. But how do I know that there is more than this "idea" or "image" before my mind's eye? I might try to confirm that the apple is real by reaching out to it. But as I touch it, in the mediationist scenario, I am still not directly in touch with the real apple but only with tactile sensations, ideas, or representations. The question that was raised initially with respect to visual mediators applies equally to tactile mediators, and to any other sense I might appeal to. There simply seems to be no way to get out of the circle of mediators, once these are understood as entities inside the mind.

Perhaps the notion of resemblance—as used by Locke—can come to the rescue. Some of our ideas resemble objects and thereby warrant contact with the world. But, as George Berkeley pointed out, this raises many new and quite difficult problems (Berkeley, 1710). Berkeley argued that it made sense to hold that one mental entity can resemble another mental entity. My impression of pink can resemble my impression of red. But it remains unclear what it would mean, argues Berkeley, for a mental entity to resemble something nonmental.

Berkeley himself shared many of the assumptions of the philosophy of mind of Locke and Hume. But because of the problems of mediationism, he concluded that we don't access a world beyond our ideas at all. The mind remains confined to the mind. All we ever do is experience mental entities. Of course, the order in which we experience these is not haphazard. If we experience the visual ideas of seeing an apple, and we reach out to touch it, we will consequently experience tactile ideas. But such ideas are not produced by a real, external apple that first caused the visual ideas and later the tactile ones. Rather than by external objects, the order between our ideas is produced by God. It is God, not the external world, who weaves the patterns present in our experience, and on the basis of which we believe in an external world.

Berkeley was an idealist. Everything, including what we thought of as the physical world, proves, on further reflection, to be mental. Such idealism has always been considered as a radical position, and one to be avoided. Berkeley argued his idealism logically follows from mediationism. How can mediationists resist having to follow Berkeley into idealism?

They can try to do so in several ways. One can provide a theological solution, à la Descartes, and posit that God could not allow us to be so deceived as to falsely believe in the existence of an outside world. There are less idiosyncratic ways to argue that, versus Berkeley, there is a way to get from internal ideas to external objects. One option is to invoke causation and argue that we somehow can know with confidence or justification that our ideas are caused by an external world. Or one can provide non-causation-based theories of how internal representations manage to make contact "from within" to the external world. One can, in contemporary idiom, provide a theory of content for mental representations. Theories of content will be discussed in chapter 9.

3.4.2 Sensations

Both according to philosophers such as Locke and Hume and scientific psychologists, there is an important difference between simple and complex experiences. Hence, they talk of simple ideas and impressions, such as color experiences and complex ideas or impressions, such as the idea or impression of a cat. Moreover, Locke describes in some detail mental operations by which he thinks complex ideas are formed out of simple ideas. Note that one finds such thinking also in Descartes, for example, when he explains how seeing distance results from a process that considers the tactile feelings coming from the muscles of each eye (see chapter 2, 2.3.3). Therefore, the idea that perception involves a causal process from a stage involving simple experiences to a stage involving complex experiences was already there when psychophysics arrived on the scene. By the time psychophysics took off, the simple/complex distinction had become one between sensation and perception. Moreover, psychophysicists often very explicitly saw this distinction reflected in a causal process. Sensations come first and cause perception.

Critics have claimed that there is no evidence for such prior sensations. This point was made at various times throughout history, for example, by Thomas Reid (1710–1796) and Samuel Baily (1791–1870). Reid was a contemporary of Hume, whose influential ideas he strongly opposed. In two of his major works (Reid, 1764/2000 and 1785/2002), Reid developed argument after argument against a philosophy of mind, perception, and cognition in the spirit of Locke and Hume. First, Reid agrees with Berkeley that if perception is mediated by ideas or impressions, one never perceives the external world. But instead of following Berkeley to conclude that, therefore, the external world does not exist, Reid comes to the opposite judgment that, therefore, impressions and ideas don't exist (Reid, 1785/2002, Essay II, chapter X). Further, Reid argues there is no evidence whatsoever for mediationism. That is, when we see a cat or a tree, we just immediately see the cat or tree, without any intervening stage of perceiving sensations, impressions or whatever.

Reid is interested in sensation, though (Reid, 1785/2002, Essay II, chapter XVI). He insists that sensation is not a first stage prior to perception. Reid's rejection of the existence of a prior sensational stage from which perception derives doesn't amount to a rejection of the existence of sensation. What Reid objects to is that sensations are independent from perception and occur before it. Rather, in Reid's view, sensation and perception occur simultaneously. In other words, Reid denies that seeing an apple is the causal effect of apple-related sensations in one or more modalities, yet he holds that we see apples and that the seeing comes with experiencing sensations. In fact, Reid is often credited with being the first to use the words "sensation" and "perception" to theorize about the mind. An often-cited passage in this respect is the following:

> The external senses have a double province—to make us feel, and to make us perceive. They furnish us with a variety of sensations, some pleasant, others painful, and others indifferent; at the same time, they give us a conception and an invincible belief of the existence of external objects. (Reid, 1785/2002, II xvii, p. 265)

In Reid's view, when we are, for example, looking at a rose, we have both sensations of its color and smell and a perceptual experience of the rose-out-there. But, importantly, in his view we perceive the rose directly: We don't perceive the sensations first and by perceiving the sensations perceive the rose. Indeed, in his view the sensations are quite unlike things perceived. Sensations are ways in which the mind is affected that correlate with ways in which the body is affected. Yet sensations, in Reid's view, are not themselves perceived. They are had, or felt. Having or feeling pain is not perceiving an inner painful object, but it is "being pained" (Reid, 1785/2002, pp. 35–36). In perceiving something, there is a distinction between the thing perceived and the act of perception, says Reid, but in feeling pain, there is no such distinction. Pain exists when it is felt, and when it is felt it exists, but it makes no sense to speak of an unfelt pain. In contrast, the rose one sees doesn't cease to exist when one closes one's eyes. According to Reid, sensations deriving from the use of our external sensory organs, such as sensations of color, are like pain. They too are not perceived but felt. Hence, on Reid's view, it makes no sense to think of perception as being of sensations first or directly, or that by perceiving sensations directly one perceives the outside world indirectly (see also Ben-Zeev, 1986). According to Reid, sensation and perception happen simultaneously. When we open our eyes in front of a rose, we at the same time see the rose before us and are affected by its colors. Crucially, we don't see the rose *because* we perceive its colors first. Indeed, we might not pay attention to our sensations and perceive, for example, the hardness of a table without the accompanying tactile sensations (Reid, 1785/2002, pp. 195–196).

Perhaps Reid's critique that there is no evidence for a prior stage of sensation can be countered by assuming that sensations can be thought of as unconscious. Perhaps there is indeed no evidence for a stage of conscious sensation occurring before perception, but why not modify the theory and take the sensations to be unconscious? Exactly this move was made in history. While in its earliest stages psychophysics studied conscious sensation, by the time Helmholtz wrote his influential works on perception, sensations, as well as their processing, had become unconscious. As mentioned, Helmholtz's notion of "unconscious inference" was proposed for the first time by John Stuart Mill (1806–1873), in his *System of Logic* (Mill, 1843). Helmholtz had read Mill's work (Reed, 1997). Mill was a card-carrying empiricist, claiming that "sensations are . . . merely colors variously arranged" (Mill, 1842/1864, p. 167; Reed, 1997, p. 138). To arrive at perception, these need to be worked upon by mental processes. Mill was the first thinker to conceive of these operations as internal, unconscious inferences. They were similar to logical reasoning, as the classical example of a syllogism. In such syllogism, one reasons from premises such as "Socrates is a human" and "All humans are mortal" to "Socrates is mortal." Mill modeled the process mediating between sensation and perception on such an inference. Sensations formed the premises of this process. But this process, including its premises—sensations—was held to be unconscious. This protected the theory from the objection that we don't experience ourselves engaging in inference when we are perceiving—or at least not in standard circumstances.

At first glance, this might look good. But notice that a problem arises to make this sit well with the motivation for the overall mediationist framework. An important motivation for the mediationist program was to analyze what we know, what's in the mind by building it up from the immediately given. Thought and perception were not to be taken for granted but attempted to be grounded in something firm. Simple ideas, impressions of sensation, and psychophysical sensations were called to duty to perform exactly this foundational role. This all seems very plausible and intuitive. When you experience seeing a cat, you might be mistaken about there being an actual cat. It seems to you that you are seeing it, but you might not be seeing it. On a dark night, you might mistake a shadow for a cat. But certainly, you are having sensations. You can't be mistaken about the fact that you are experiencing sensations in the way you can be mistaken about cats. In mediationist philosophy, in psychophysics and in standard neuroscience of perception, it is the relation between simple mental entities and what's derived from them—complex ideas, percepts, thoughts—that is problematized. It was from Locke's simple ideas, Hume's impressions from sensations, and the psychophysicist's elementary sensations that the rest of the mind was built up by various processes, such as association, and later inference.

Yet, if sensations are what's more obviously given to us than perception, how can they be moved out of consciousness? This criticism was at the heart of a back-and-forth between Samuel Bailey and John Stuart Mill (Bailey, 1842; the case is discussed in Reed, 1997). Bailey argued that the very idea of unconscious sensations made no sense within the overall mediationist assumptions made by Mill. That is, Mill's starting point—one shared by many in the tradition he is part of—is that we have certainty about sensation but not about perception. We know for sure about colored blobs and sensations of touch but not about external objects. That is why the tradition proposes a theory is needed to explain how we get from (certainty of) blobs and other sensations to perceptions. Against this background, making the sensations unconscious becomes problematic. Then one takes away what one first assumed, namely, the certainty about blobs and touches. They become unconscious. Only perception is conscious. But then perception becomes more certain than blobs and other sensations. A contradiction appears inside the overall system.

3.4.3 Perceptual Error and Imagination

Sometimes, so many thinkers about the mind have claimed, we seem to be perceptually experiencing something, but we aren't. You can think you see a cat suddenly appearing in the dark alley, but when you get nearer you realize there's only a shadow. At night you might dream of seeing a dragon and visually inspect its scales to see that they form a pattern of gold and silver. Or you might imagine such dragon and see it before your mind's eye. Isn't it the case that in all these experiences, though you do not really perceive something, as there is nothing real to perceive, you have experiences that are very similar to perceptual experiences? And if so, doesn't that mean that perceptual experience must be mediated by something standing between you and the outside world?

That is, mustn't these special experiences involve mediation of some sort? You cannot be experiencing what is there, but still are experiencing something, so mustn't there be something else that you experience—something like an image, idea, or representation of a cat or dragon? The reasoning can then move on. Because experience in these special circumstances feels so similar to experience in actual perception, could what explains perception also be something like an image, an idea, or a representation? In short, shouldn't it be concluded from cases of perception-like experience that perception is mediated?

One way to keep this argument from taking off is to reject that the kind of perception-like experiences pointed at need mediators in the first place. Why say that seeming to perceive a cat requires a mediator? For an alternative explanation—one that doesn't involve mediators—we can again turn to Thomas Reid. Reid is aware that we can seem to perceive, while we don't. Yet, he explains seeming perception without invoking mediators. His starting point is that in most circumstances, perception works perfectly well. One only sees a cat in the alley when there is a cat in the alley. But this doesn't mean that perception always operates reliably. In some circumstances normal functioning is disrupted or just doesn't manage to deliver. If there's damage to one's eyes or one's optic nerve is compromised, one might see two objects where there is only one. If the circumstances are different but similar to circumstances in which perception would function effectively, one might seem to see things that are not there. Reid refers to effects skillful artists bring about. In a trompe l'oeil, they make spectators have the experience of a three-dimensional object, while in reality there's only paint on a white surface. The skillful artist can achieve this effect by creating conditions closely resembling those in which the seemingly seen object would be present. What they paint on the wall affects the perceiver's eye in ways very similar to how the real object would affect the perceiver's eye (Reid, 1785/2002, Essay II, chapter 22). Crucially, the explanation of seeming perception offered by Reid doesn't invoke a mediator. The explanatory burden is carried by the circumstances, both those in which the seeing operates well and the closely similar conditions in which it doesn't.

Interestingly, Reid explicitly defends a mediator-free account of fictional imagination. Suppose, to iterate his example, that you image a centaur. Isn't it undeniable that in this case one must construe what you do as being related to some mediating image, idea, or representation of a centaur? No, says Reid, when you imagine a centaur, you are imagining a centaur, not an image, idea, or representation of a centaur (Reid, 1785/2002, part IV, chapter II). What you are imagining is alive. It has a head and breathes. Ideas, images, and representations don't do that. Reid doesn't deny that there are no centaurs, but he insists that you don't need an imaginary mental centaur-like object to make sense of imagining a nonexisting centaur. Reid insists on a distinction between the act of imagining and the object of imagination. The act is real, but as its object isn't, it shouldn't be thought of as some real—mediating—object. According to Reid, when we say that imagination has an object, we only say that imagination is about something. To say that the object of someone's imagination is a centaur is the same as saying that someone is imagining a centaur. Imagining a centaur itself doesn't logically compel one to assume that an image, idea, or representation is involved. Reid agrees that one might say such things as "I have the image of a centaur." But saying that, is just saying that one imagines a centaur. To "have an image in the mind" and "to imagine" are different ways of speaking about one's act of imagining. On Reid's account it is a serious mistake to interpret the phrase "having an image in one's mind" literally. He writes: "But to infer . . . that there really is an image in the mind, distinct from the operation of conceiving the object, is to be misled by an analogical expression; as if, from the phrases of deliberating and balancing things in the mind, we should infer that there really is a balance existing in the mind for weighting motives and arguments" (Reid, 1785/2002, p. 322).

3.4.4 Turning the Tables

Consider again the passage in which Hume gives the example of the perception of a table to motivate a mediationist picture of perception. Hume writes: "Nothing can ever be present to the mind but an image or perception" (Hume, 1748, 12.9, p. 114). And further: "The table, which we see, seems to diminish, as we remove farther from it: but the real table, which exists independent of us, suffers no alteration: it was, therefore, nothing but its image, which was present to the mind. These are the obvious dictates of reason" (Hume, 1748, 12.9, p. 114).

One way in which Reid takes issue with Hume (Reid, 1764, 1785/2002) is by considering in detail the perceptual situations Hume rather casually refers to. Reid shows that the projection on the eye changes systematically and predictably with the position of the observer. Think of it this way. Hume says you don't experience the table itself, but an internal image. Why? Because something changes with observer position. Then he concludes that it must be an inner image. What Reid shows is that something perfectly outer changes in exactly the required ways: the projection of the table on the eye.

Think about taking photographs of a table with your phone from different places. At all the positions from which you can take a picture, you can do so because there is some physical image or projection of the table that can be captured by your camera. These projections become smaller when you walk away from the table. The fact that you can take pictures from many spots proves that a projection is available at many places—even if the shape of the projection changes.

Reid links this projection with what he calls "visible figure" (Reid, 1764/1997, p. 79). The visible figure is how an object is present to a visual perceiver from a certain position. Visible figure correlates with projection. As its projection does, the visible figure of a large object shrinks—in predictable ways—as its distance to a visual perceiver increases. This visible figure plays the role that was pointed at by Hume as requiring a mental image. It becomes smaller when one occupies a place farther from the table. The rules for projection are well known in geometry. Reid stresses that the way projections and their correlated visual figures change is perfectly objective, but depends on both the objective table and the objective position that a perceiver occupies with respect to the table. It is relational. According to Reid, it is because we are sensitive to the changing visible figure that we see the actual size of the table. Or because we are sensitive to the diminishing size of the visual figure of the table, we see the table as having a constant actual size. This sensitivity to visual figure thus delivers what's needed for an understanding of how perceivers are in visual perception in contact with reality without invoking all the troubles that mediationism incurs. Reid turns the example Hume gives for mediationism into one *against* mediationism by noticing that the account in terms of visible figure offers something that Hume's imagistic account doesn't. Reid's external and objective story explains how visible figure changes with position. What explains changing visible figure is simply the geometry of projection. In contrast, Hume cannot explain why the mental images, when walking away from the table, change. In Reid's own words: "Add to this, that upon the contrary hypothesis, to wit, that the objects of sight are internal, no account can be given of any one of those appearances, nor any physical cause assigned why a visible object should, in any one case, have one apparent figure and magnitude rather than another" (Reid, 1785/2002, p. 183).

As we will see in chapter 10, Reid's resolution of the table challenge exhibits a striking resemblance to a research program in perceptual psychology known as "the ecological approach" to perception. Reid further claimed that the sensitivity to the visible figure of objects can be acquired. By visually interacting with objects, you get acquainted with the different visible figures connected to them and so come to see their actual shape. Again, this idea will return in the ecological approach to perception, as well as in the sensorimotor contingency theory of perception, which will be presented in chapters 10 and 11.

3.4.5 Reification

Reid's thoughts on the acquired nature of shape perception can be developed further. Suppose we are in visual contact with the table by building up a sensitivity to the ways in which its appearance systematically changes with our position with respect to the table. By acquiring this sensitivity, we acquire a grip on the objective table. Note that in the reasoning here, this grip is the *result* of a process of getting acquainted with the table's appearances. We don't have an idea of an objective table first but develop our sensitivity to the objective table along the way. In Reid's philosophy of perception, seeing the table comes along in the activity of the person. It is not a thing inside the person, let alone a thing that guides the person's visual exploration.

It is possible to think of perception as deriving from a "percept" inside the person. But if perception is not a thing, such thinking is a *reification*. To reify is to make into a thing what isn't a thing. Democracy isn't a thing, in the sense that one cannot locate it as one locates a parked bicycle. Without him using the term, many of Reid's arguments are aimed at criticizing reification when thinking about the mind.

The case against reification forms an important thread in the work of the American pragmatist philosopher John Dewey (1859–1952). He diagnosed much theoretical thinking about the mind as involving reification. He gave a name to this the tendency: "the historical fallacy." Dewey sees reification at work in the notion of sensation, which was so important to modern philosophical and early psychological thinkers.

To see this, we need to turn to Dewey's famous paper "The Reflex Arc Concept in Psychology." There he offered a critique of the psychological notions of stimulus and response. Dewey gave a wider interpretative framework for the problems that were raised for sensations by Thomas Reid (Dewey, 1896).

Dewey notes that sensations occur within ongoing, purposeful activity of organisms. Organisms seek stimulation and actively move and optimize their position to receive stimulation that might lead to feeling and perceiving. Something is sensed, but what is sensed is dependent on the organism's current and intended activity. To illustrate this point, Dewey writes: "If one is reading a book, if one is hunting, if one is watching in a dark place on a lonely night, if one is performing a chemical experiment, in each case, the noise has a very different psychical value; it is a different experience" (Dewey, 1896, p. 361).

The main thesis of Dewey's article is that one cannot consider stimuli and responses, and hence, sensations and movements of an organism as self-standing entities. These concepts only make sense when one recognizes that they are aspects of an organism's ongoing adaptative activity. He gives the example of a child in front of a burning candle. The child sees a light, reaches for it, and gets burned. On Dewey's account, the actions that the child undertakes determine the sensations they have. In the first place, the child has to look to see the light. More fundamentally, there is never a neutral "seeing of a light." Instead, there are light-for-reaching sensations that transform into light-that-burns sensations. The activity determines the sensation of seeing. But similarly, the reaching is a reaching to the reachable light, and the withdrawal coming with the burn is a withdrawal-from-a-burning-light. Just as the visual sensation is "constituted" (Dewey, 1896, p. 363) by the motor act, so is the motor act constituted by the visual sensation. Stimulus and response, sensation and motor reaction are inseparably related to a broader organismic process of coordination.

Dewey notes that it is crucial, when considering stimuli and responses, to make a distinction between unconscious and conscious (re-)actions. When activity with respect to a certain stably reappearing situation has become stereotyped—as when typing a text on a keyboard for a skilled typist—the cooperation is fixed, and one can distinguish clearly between the stimulus aspect and the response aspect. But when the activity does not have this stereotyped character, when the coordination between the organism and its environment still has to be forged or established, what is sensed as stimulus only gets resolved once the action-in-response is formed and executed.

Dewey warns for what he terms "the historical fallacy" (Dewey, 1896, p. 367). This is, first, to take what is the result of the process of coordination, namely, the resulting sensation and the resulting activity, not as aspects that are abstracted from an encompassing coordination process but as independent existents. The fallacy, as described by Dewey, is exacerbated by treating these alleged entities as the *causes* of the coordination. So, first there is the making of the result into a thing, and then the imaginary thing is taken to be a cause of the process that led to this result. Committing the fallacy means turning things on their head. The fallacy construes what exists only inside a process as something that can be isolated from it and can occur on its own—a sensation constituted by and inside a cooperation process becomes the source or cause of that process.

Dewey's thought thus fully concurs with the critique of sensations as causes of perception, which was aired by Reid, Bailey, and other thinkers. Though by no means did it become widely accepted, this critical view never died. Nowadays it remains a main motivation for active research programs, as we will see in chapter 10. The founder of one such program, ecological psychology, J. J. Gibson formulated a very Reidian conclusion, when he stated, "From all the[se] facts of vision, hearing, and touch . . . we ought to conclude that sensations are not the cause of perceptions" (Gibson, 1963, p. 8).

Bibliographical Note for Chapter 3

There are many, many texts on the history of psychology. One with an interesting philosophical angle is Edward Reed's *From Soul to Mind. The Emergence of Psychology from Erasmus Darwin to William James* (Reed, 1997). The line of the current book is congruent with Reed's, as the references in this chapter show. Reed has much more to offer than can be covered here. Particularly interesting is his treatment of the impact of views of the mind presented outside of academia, for example, in novels. He also proposes the intriguing view that psychology didn't grow out of philosophy, but rather the other way around. That is, Locke or Hume, for example, could be considered with equal right to be pychologists as philosophers. Philosophy-as-we-know-it only became a separate discipline after psychology had emerged as a self-standing institutional discipline, as Reed argues in his chapter 10. Reed also shows that the early psychophysicists were much more than hard-nosed scientists. For example, he details the metaphysical speculations of Fechner, on which the German thinker published prolifically (see Reed's chapter 5). Graham Richards (1992) takes a line opposite to Reed, arguing that Locke and Hume were more philosophers than psychologists.

Cobb (2020) contains a well-informed and often refreshingly critical account of the emergence and development of thinking about the brain.

Contemporary philosophers endorsing a representational view of perception and mind will want to point to differences between their views and the mediationist line of thinking discussed in this chapter. Chapter 9 of this book will offer a bit more on current philosophical representationalism about perception. For a comparative exposition and defense of contemporary representationalism, see Bayne (2022), chapter 4.

An excellent contemporary book on Thomas Reid's views on perception is Nichols (2007). The reader should be aware that throughout the book, Nichols uses the terminology that is currently standard in philosophy of mind ("mental states," "propositional attitudes"), but was nonexistent in Reid's times and not to be found in Reid's own writings. Talking of which, Reid's *Essays on the Intellectual Powers of Man*, published in 1785, when Reid was seventy-five, remains highly readable and remarkably fresh.

Reification is explored and applied in a Deweyan spirit in van Dijk (2016, 2021) and analyzed and criticized as a persistent problem in cognitive science in Zahnoun (2020) and Iani (2024).

4 Evolution: From Biology to Behavior, Cognition, and Culture

> Darwin's idea . . . eats through just about every traditional concept, and leaves in its wake a revolutionized world-view, with most of the original landmarks still recognizable, but transformed in fundamental ways.
>
> Darwin's idea had been born as an answer to questions in biology, but it threatened to leak out, offering answers—welcome or not—to questions in cosmology (going in one direction) and psychology (going in the other direction).
>
> —Daniel Dennett (1995, p. 63)

4.1 Evolution in Mind

Why bother with evolution when talking about the mind? One could think that evolution is primarily a biological subject. Evolution, so it might appear, explains why there are different species of animals and why they have the bodies they have. But the growth of the "tree of life" is a process in the long term—the very long term, given the first available evidence for living beings is dated back 3.7 billion years. The case for involving evolution might even look weaker when it concerns human minds. For haven't we humans created an environment in which we are shielded from the effects of evolution? In nature, life might be a matter of "survival of the fittest," but not in contemporary societies—or so thought Bertrand Russell (Popp, 2007, pp. 5–8).

There are several reasons to think that such sidelining of evolution when talking about the mind should be avoided. In the first place, recognizing that minds have evolved allows for a fuller understanding of them. It permits us to see how minds of different species are related to each other. Tracing down the natural origins of minds, or where they come from, might lead to an enhanced comprehension of what they are like now—just like getting acquainted with a person's youth might enhance insight into their current actions and personality. Further, the kind of processes that occur on a large timescale in biological evolution and give rise to the origin and divergence of species—phylogeny—might reappear on smaller timescales. Exploratory activity—trying things out and retaining what works—looks like having a similar structure to evolution that gives rise to phylogenetic diversity. Cultures evolve too, and increasingly, theorists are arguing that cultural change should be understood in evolutionary terms (Boyd, 2018; Boyd & Richerson, 1985, Henrich, 2016; Laland, 2017; Skinner, 1981).

In chapter 2 we saw that, with few exceptions, the mechanistic thinkers of the seventeenth century placed strict restrictions on the reach of mechanistic theorizing. The mind was off-limits, as was, for most of these thinkers, life. The goal-directedness of life, and the operations of the rational mind, were, for the typical seventeenth-century thinker, obviously unexplainable on mechanistic grounds. As science was seen as inherently mechanistic, life and mind were placed outside of science. These subjects could not be exhaustively comprehended by applying the scientific methods appropriate for inanimate nature but required invoking what stands outside nature.

From its beginnings, Darwinian evolutionary thinking was aimed at providing a naturalistic account of the goal-directedness of animate nature. The marvelous ways in which the parts of the body function to allow animals to live and thrive, when seen as products of biological evolution, emerge from within nature, instead of requiring a source outside of nature. If evolution has the potential to naturalize the recalcitrant feature of animate goal-directedness, perhaps it can also allow to naturalize the features of the mind that have been thought to lie beyond the reach of science, features such as rationality, abstraction, meaningful talking and thinking, or sensory consciousness. Evolution has acted like a magnet for many thinkers who have tried to get a grip on minds and on its allegedly intractable properties, as we'll see further on (see especially chapters 9 and 11).

But let us start with Darwin.

4.2 Evolution by Natural Selection: Core Ideas

The idea that biological life has arisen and has acquired its variety through a process of evolution is now a widely accepted part of a naturalistic, scientific worldview. Its core idea is that because of biological evolution, the tree of life has grown and diversified. Charles Darwin (1809–1882) is universally recognized for providing the first thorough and systematic theory of biological evolution, much of which still holds. Darwin published his theory of evolution in his celebrated *On the Origin of Species by Means of Natural Selection, or the Preservation of Species in the Struggle for Life* (Darwin, 1859).

In its original guise, the theory of natural selection is a theory about how species of living organisms evolve. A species is a group of animals that interbreed, or of which the members can have fertile offspring. The core ingredients of the original theory are *variation, differential fitness,* and *heredity* (Lewontin, 1970).

Variation refers to the fact that existing members of a species differ in their characteristics. For example, some might be bigger, faster, or more perceptive than others. Varieties of characteristics might be better suited to a particular environment. Being bigger might be an advantage when it comes to fighting, but a disadvantage if one has to walk high up on twigs that bend and drop you off if you're too weighty. If one's diet mainly consists of the leaves of tall trees, in an appropriate environment, animals with longer necks will have more food available than animals with shorter necks. So, if there's variability of characteristics there will be *differential fitness:* differences in how well organisms thrive in their environments. Some individuals with certain characteristics will do better than other individuals.

If the offspring of the organisms with favorable traits tend to have these traits as well, *heredity* is at work. Given variation, differential fitness, and heredity, in the long run the number of individuals with those characteristics will increase in the total population of individuals of the species. Inheritance is key to this process of natural selection of characteristics, because if traits were not passed on to the next generation, advantages would die with the individual that held them. Organisms with properties that are in some way adapted to the environment tend to be favored in this intergenerational process and species will gradually evolve to become more adapted to their environments. This is *evolution by natural selection.*

It was important to Darwin and many subsequent evolutionary theorists that adaptation was not understood teleologically, which means as guided by or as directly responding to a need from the environment. Variation in properties of organisms was taken not to arise *because* some variants better fulfill some function than other variants. Rather, variation occurs for independent reasons and some variants just happen to better fit the environment than others. These variants are retained, or selected, and become more common in a population because having these properties is transferred from parents to children. To illustrate, some giraffes just happen to have longer necks. These giraffes are "fitter": They can obtain more nutrition, and as a result, they will be in a better position to leave progeny. If the progeny inherits the properties of the parents, over time, giraffes with long necks will become more common. And then necks can become even longer because this process can repeat itself.

Darwin had no idea of what exactly made variation and inheritance possible. But later developments in genetics, initially due to Gregor Mendel (1822–1884), allowed evolutionary theorists to fill that gap. As a result, in the first half of the twentieth century, the so-called "modern synthesis" emerged and became the established understanding of evolution (Huxley, 1942). Its genetic component was deepened by the formulation of the DNA double-helix structure of genes deriving from the work of Rosalind Franklin, Maurice Wilkins, James Watson, and Francis Crick roughly halfway the twentieth century (Pray, 2008).

The main lines of the modern synthesis are very Darwinian: Biological evolution is an adaptive process, due to natural selection of traits. Traits vary in individuals. Those individuals with traits that better fit the environment are better able to procreate, so that, over generations, the "better" traits become more common and organisms become better adapted to their environments. Genes play a central role in the modern synthesis, because genes are the basis of inheritance, of development, and of change. Traits are passed on from generation to generation because genes are passed on. Genes are thought of as the "blueprint" containing the necessary information or instructions for how an organism develops from a single cell into its adult form (Lala et al., 2024, p. 269–270). Moreover, genes are the locus of changes in traits. Traits change because genes change, so different traits will arise when genes are modified. As in Darwin, according to standard modern synthesis theorizing, these changes in genes are unrelated to how well the trait that results from the change fits the environment. Instead, random changes happen in genes, for example, as a result of being exposed to too much sunlight, and trait differences result. Most random changes–mutations of genes–do not lead to more adaptation, but some mutations do. Evolution toward adaptation occurs by the accumulation of "lucky" mutations.

Organisms become more adapted to the environment and organisms that are less well adapted are filtered out.

If organisms and their parts beg to be interpreted in teleological terms, or in terms of goals that are served, the theory of evolution by natural selection offers a means to explain this teleology by tracing its origins to nonteleological processes of variation and selection. At the bottom of all this functionality stand aimless processes.

Ever since its inception, evolution has faced the question of how possibly evolution by natural selection could explain complex structures such as the eye. Eyes have very intricate internal parts, which need to be minutely orchestrated with each other for the eye to function. Moreover, the whole eye is anatomically and functionally embedded in a body, and intense and very fine-tuned cooperation is necessary between muscles, nutrients, and the brain for seeing to be possible in actual circumstances. Both Darwin and theorists of the modern synthesis have answered that such complexity can arise by accumulation of small, gradual changes. They have been committed to *gradualism*: There are no jumps in evolution, no sudden emergence of large changes. If large-scale changes occur during evolution, these happen because of the accumulation of many small changes (Lala et al., 2024, p. 173–174).

4.3 Evolution of Action and Intelligence: Darwin's Earthworms

Evolution aims to explain the emergence of bodies and finely tuned anatomical features. But living organisms are more than coordinated structures. They develop, grow, and move in environments. They act in adaptive ways: ways that enhance their existence and procreation. The spotlight of evolutionary thinking can be directed on organismic activities, just as it can be on organismic bodily and physiological structure.

Indeed, Charles Darwin investigated with great dedication and care animal activity. One group of animals to which he devoted much attention in his later work was earthworms (Darwin, 1881). Darwin was intrigued by the question whether they could be considered to have "mental qualities" or "mental powers." Based on observation and experiment, carried out over many years, Darwin concluded that the activities shown by earthworms proved they indeed deserved to be ascribed mental qualities and powers. He argued that the worms can direct their attention, and depending on it, become more or less sensitive to the same stimulation. Darwin claimed earthworms show some form of awareness. They are timid, differentially enjoy eating different foods, and overcome their dread of light when their sexual passion is strong (Darwin, 1881, pp. 48–49). Moreover, they act intelligently. In various of their activities, they chose among a variety of possible actions those that were most fitting to the circumstances and changed their strategies as circumstances changed. Crucially, such adaptive behavior occurred spontaneously and was not the result of long series of trial and error. Hence Darwin writes: "One of their strongest instincts is the plugging up the mouths of their burrows with various objects; and very young worms act in this manner. But some degree of intelligence appears . . . a result which has surprised me more than anything else in regard to worms" (Darwin, 1881, p. 50).

What led to Darwin's surprise? To begin with, earthworms don't have specialized sensory organs like eyes or ears. Nor do they have a brain. Yet they are sensitive over their whole skin, which can be considered as a distributed sensory organ if one wishes (Reed, 1996, pp. 20–21). Earthworms burrow holes in the ground to protect themselves from desiccation and heat and for other ends. At the entrance of the hole is a slightly wider basket-shaped area, that the worms plug up with materials they find in their environment, such as leaves or twigs. Darwin did revealing experiments on how the earthworms handled leaves.

The best way to pull a leave into the entrance so that it is sealed off, is by grasping it by its smallest end and dragging it in the burrow by that side first. Most leaves are at their smallest at their tip, to become broader at the base where the petiole is. In many circumstances, earthworms follow this strategy. This already shows some perceptual capacity or "power" as Darwin writes. A simple explanation of the worms using this strategy would be to assume it is due to some hardwired rule to the effect that leaves must be grasped at the tip, opposite the petiole. But such simple explanation can be discarded, because the earthworms will adapt when they are given leaves—Darwin used those of a rhododendron—that are small at the petiole and broader at the tip. The worms tended to take many of such leaves at the small side, at the petiole. They also chose this efficient way of acting frequently when presented with leaves Darwin cut from paper. Moreover, the worms didn't need to learn the better strategy by an extended process of trial and error. They switched spontaneously to the advantageous way of action when different leaves were presented. By cleverly ensuring that his cut-paper leaves would show traces of worm activity, Darwin was able to show the earthworms did not initially attempt to pull the paper leaves in by their broad ends (Darwin, 1881, p. 116). Further, the earthworms were sensitive to the overall circumstances, to adopt the most efficient strategy when a first one led to failure. When Darwin made the air in a terrarium warm and moist, the worms weren't very considerate in how they pulled leaves. That makes sense, because in such conditions, the disadvantages of not having the burrow plugged up almost disappear. When the air was made colder and dryer, the worms adopted the more careful ways of handling leaves.

These observations, alongside many others, made Darwin conclude that earthworms have mental qualities and powers. Darwin ascribes intelligence to them because they acquire "some notion, however rude, of the shape of an object and of their burrows" and they "act in nearly the same manner as would a man under similar circumstances" (Darwin, 1881, p. 116).

Darwin's attitude toward the intelligence of animals as humble as the earthworm differs markedly from that shown by Descartes. Instead of describing their activity as machinelike, he notices its adaptedness and sensitivity to circumstances. Darwin sees intelligence in the capacity to flexibly switch to actions that are appropriate to the sensed environment. Even if Darwin sees the worms as "having notions," he immediately adds these are "rude"—a far cry from Descartes's clear and distinct ideas.

Moreover, if earthworms are considered intelligent from a Darwinian perspective, then intelligence must be both biological and evolved. Intelligence then becomes explainable via evolution. Darwin's works on earthworms hints at the potential for an evolutionary account of mind: an evolutionary psychology. Darwin's own evolutionary psychology of earthworm intelligence brings to the foreground interaction with the environment: how organisms act on their environment and the environment on organisms. But stressing interaction was not what mainstream psychology was doing, where we left it in the previous chapter.

4.4 Evolution in Psychology: Selection by Consequences

Think again about the emergence of the science of psychology through psychophysics. The founding fathers of psychology studied perception in abstraction from organismic action. They were interested in how physical stimuli gave rise to internal activations of nerves and to sensations. Fechner (1860) drew a contrast between "outer" and "inner psychophysics."

"Outer psychophysics" concerned relations of stimuli to sensations. "Inner psychophysics" concerned the relation of innervation to feeling. So though "outer" referred to something outside of the body and mind, it only referred to reception of stimulation from the world. What was left out of the picture was how the organism affects the world.

The one-sidedness of the psychology of sensation and perception was criticized by Dewey (see chapter 3, section 3.4.5). He pointed out various ways in which organismic action *does* matter. In the first place, the environment is not neutral. Not everything counts equally for organisms with different needs, different bodies, and different capacities for action. There are no neutral stimuli: What an organism can do in light of its needs matters. Moreover, very concretely, whether and how an organism is acting *now* determines what it perceives. The same touch, sound, or light is experienced very differently in different circumstances, depending on whether the organism is passive or active, on what movements it is making, and what actions it is engaged in. Whether or not something becomes a stimulus, and what stimulus it will become, depends very much on the organism's activities.

So, actions determine and change perception. But one can ask the further question: How do actions themselves change? And one can give an evolutionary answer: by selection of actions, because of their relation to the environment.

At some point, action, its evolution, and selection did come to the forefront in some strands of psychological thinking. Edward Thorndike (1874–1949) was concerned with the question of how actions became better tuned to environmental situations. In 1911, he formulated "the law of effect," also known as Thorndike's law. With it, evolutionary, selective thinking is applied to action: Actions that have "good" effects in some circumstances will tend to be repeated in similar circumstances, and the opposite for actions with "bad" effects:

> The Law of Effect is that: Of several responses made to the same situation, those which are accompanied or closely followed by satisfaction to the animal will, other things being equal, be more firmly connected with the situation, so that, when it [the situation] recurs, they [the responses] will be more likely to recur; those which are accompanied or closely followed by discomfort to the animal will, other things being equal, have their connections with that situation weakened, so that, when it recurs, they will be less likely to occur. The greater the satisfaction or discomfort, the greater the strengthening or weakening of the bond. (Thorndike, 1911, p. 244)

The effects or consequences of action, and their selection, formed the core of Burrhus Frederic, or B. F., Skinner's (1904–1990) "radical behaviorism." Skinner's work is standardly inscribed in a tradition called "behaviorism." Though Skinner often described himself as a behaviorist, his theory differed considerably from the kind of theory that was known as behaviorism before he came on the scene (Chiesa, 1994; Lee, 1988). Much pre-Skinner behaviorism, such as famously developed by John B. Watson (1878–1958), focused on stimulus–response relationships. This behaviorism was influenced by the Noble Prize–winning work of Ivan Pavlov (1849–1936) on conditioning. Pavlov showed, for example, that it was possible to make dogs salivate, a reaction they show when they perceive food, to the sound of a bell. Pavlov brought this about by consistently sounding the bell before the food was presented and then sounding only the bell. In this work on "classical conditioning," the response is a fixed, evolutionary established reaction: salivating. This reaction, normally triggered by food, is now conditioned to become triggered by some other stimulus. The central scheme is fixed response, learned stimulus. What is manipulated is the stimulus.

This kind of stimulus–response behaviorism was developed extensively by John Watson (1878–1958). He used classical conditioning, for example, to induce fear in a child for an object not feared before, in experiments that were controversial in his time, and now would not be allowed for ethical reasons. Though these experiments only targeted direct stimulus–response links, later theorists in the behaviorist tradition such as Clark Hull (1884–1952) and Edward C. Tolman (1886–1959) argued that responses were not only determined by stimuli but also by what, after Tolman, came to be known as "intervening" or "mediating" variables. Clark Hull formulated the drive reduction theory, which included "drives" such as hunger or a desire to earn money as codeterminants of responses, besides external stimuli. Edward Tolman (1886–1559), argued that the behavior of rats that have learned to find their way through a maze had to be explained by means of "cognitive maps" that were formed during the learning process. Tolman wanted to replace stimulus–response behaviorism by stimulus–organism–response behaviorism (Chiesa, 1994). One way to see Hull's and Tolman's proposals is as adding a hypothetical extra "internal stimulus" to the external stimulation received by an organism. The organismic actions then become determined by a combination of the external and internal stimulation. The style of explanations aimed at in both approaches is mechanistic:

Both theorists aim to account for a current event in terms of immediately preceding events—the occurrence of external stimulation, plus eventually the occurrence of something inside the organism, be it a drive, an internal stimulus, or a cognitive map. Also, note that the alleged internal components are hypothetical entities: They are supposed to exist, without necessarily having been observed.

Skinner's approach differs in several respects from all these behaviorisms. It is concerned with a different kind of conditioning than the one studied by Pavlov and Watson, and it is not, like Hull's or Tolman's views, mechanistic. He explicitly distinguished his own approach from behaviorist theories like Watson's or Hull's, even to the extent that he at times expressed dislike of the term "behaviorism" for his own theory (Lee, 1988, p. 82).

Skinner was interested in a different kind of organismic activity than the reflex-like activities that were of concern to Pavlov and Watson. In essence, classical conditioning was aimed at adding a conditioned stimulus to a stimulus–response association that had appeared during the evolutionary history of the organism but now was fixed. The result was a new trigger for a fixed response. In contrast, Skinner was interested in organismic actions that were not initially fixed. The range of such activities is unlimited. Examples include moving from one place to another, pushing something, browsing a magazine, imagining, or speaking. Of course, not all organisms can perform all activities—people cannot dive from the sky into a lake to catch a fish, but some birds can. Organisms have their specific repertoires of activities (Skinner 1938, 1957). Repertoires can grow, but within limits.

Skinner was focused on how repertoires changed and thought the key to such changes lies in evolutionary, selective processes. In these, consequences of actions are key. What organisms do now, on Skinner's account, has to be explained by the selective history they have undergone. An organism does certain things in an environment—it moves toward food or a fire—and different actions have different consequences for the organism. Some consequences are better for the organism than others. They nourish it, guide it away from danger, or enhance its perception, rather than harm it. The selective, evolutionary assumption at the heart of Skinner's theory is that actions with beneficial consequences in the past will become more probable to be repeated in the future—very much Thorndike's law of effect. And the reverse happens if the consequences are negative. Over time, a process of selection, growth, and decline of individual activities-in-contexts occurs, similar to evolution by natural selection at the species level (Skinner, 1953, p. 90).

Skinner characterizes the core motivating idea behind his approach as "selection by consequences." Beneficial consequences were termed "reinforcers" by Skinner, and their opposite "punishments." Food is a reinforcer, burns caused by getting close to the fire are a punishment. The way reinforcement and punishment affect activities over time, Skinner called "operant conditioning" or "shaping." The actions in play are called "operants." Unlike the fixed reaction patterns studied by Pavlov, they are not pregiven. Rather than being triggered by a stimulus, operants might initially arise spontaneously and reemerge following reinforcement, as a result of the consequences they produce within a specific context. The repertoire evolves.

Skinner (1953, p. 90) describes his approach as evolutionary, but as adding a level of selected activity over and above innate behaviors: "Reflexes and other innate patterns of behavior evolve because they increase the chances of survival of the *species*. Operants grow strong because they are followed by important consequences in the life of the *individual*" (Skinner, 1981 p. 14).

One can indeed notice the presence of analogs of the three elements that were marked as defining evolution by natural selection: variation, differential fitness, and inheritance. Emitted operant activity is variable: Organisms engage in different kinds of activities in the same environment. They can achieve the same goal by different means. For example, a person can greet a friend in a great many ways, with equal efficiency. Or organisms might actively explore several possible actions in an environment. Some consequences are "fitter" than others, in that they lead to a favorable result. Selection occurs by reinforcement or punishment, depending on how well the consequences of the activities "fit." Actions that are reinforced become more probable in the future in similar circumstances; future activities are shaped by previous activities, and hence, there is an analogue to inheritance. This process of selection by consequences implies that organisms discriminate between circumstances, to be able to act similarly in similar circumstances. They learn to discern aspects of situations that are conducive to certain activities and transfer what is learned in concrete situations to novel, yet similar, circumstances (Skinner, 1953, chapter 7).

Skinner investigated operant conditioning or shaping experimentally in great detail in animals such as pigeons. He described general patterns discernable in operant conditioning. For example, he found that, in many cases, learned operant behavior needs some form of intermittent reinforcement to be maintained. He also studied the opposite of maintenance, extinction. That is, he investigated what it took to unlearn an activity that was established through operant conditioning. Skinner tried to unlearn or "extinct" shaped activities by withholding reinforcement. One of his findings was that actions that had a history of being only occasionally reinforced were much more difficult to unlearn than actions that were established by being consistently reinforced (Skinner, 1953, 70).

4.5 Operant Selectionism, Explanation, and Language

Though Skinner's operant selectionism implied a program for performing experiments, Skinner was very much concerned with defining the ends and means of psychology and therefore presented his operant selectionism as a philosophy of science for psychology. In his view, psychology should be, from the beginning, ameliorative. It should be a science that leads to improving the lives of the subjects it studies. The means psychology has to ultimately achieve this end, for Skinner, is to investigate how selection by consequences shapes activities in situations and apply such insights in the design of future environments (Skinner 1953).

The selectionist approach in psychology has crucial implications for what is considered to form explanations in the realm of the psychological and, hence, mental. How does one explain how organisms and persons function psychologically? In experimental psychophysics, but also the behaviorism of Watson, Hull, and Tolman, it is assumed that explanations in psychology are mechanistic. They must invoke regular or lawful chains of causes and effects, and effects are explained by causes. Moreover, causes typically immediately precede their effects. Skinner rejected that the framework of mechanistic causation was appropriate to understand psychological functioning. He held that to understand a psychological phenomenon is to understand how it became shaped, or how it gradually came about in a process of selection by consequences. For example, if one wants to understand and explain why a pigeon pecks at a certain lever, one has to refer to the reinforcement process that the activity was involved in: The lever provided food, a reinforcer. Such explanation of the pecking is historical. It explains what happens now by referring to events that happened in the past. A mechanistic explanation would only allow, as explanandum, or what does the explaining, what immediately precedes the pecking. Presumably, some brain process might qualify. But the selective process is temporally more distant than that. Skinner argued that explanation by invoking a selective history is a form of causal explanation too, but one different from mechanistic causal explanation (Chiesa, 1994; Skinner, 1981).

He thought that psychologists, past and contemporary, were too often tempted by the urge to provide mechanistic accounts of behavior. He warned for what he considered pseudo-explanations: explanations that were imagined to be neurophysiological and mechanistic, but in reality were pure speculation. For example, one can assume that drives (see Hull) or beliefs or knowledge about the environment (see Tolman) are identical with neural entities or processes, and invoke them, as allegedly mechanistic causes of what an organism currently does. But, according to Skinner, to construct hypothetical, imagined, neural events is only to pretend, or to imagine, to do neuroscience (Zilio, 2016). In Skinner's view, "knowledge about the environment" is not a neural process, like, for example, an increase in the electrical potential of the axon of a neuron is. Moreover, he insisted that explananda such as "a drive" or "knowledge about the environment" themselves need to be explained. Even if it is ignored that there is no established valid way of identifying knowledge with brain processes, one could always ask: But *why* is this brain process knowledge? And to answer that question, Skinner argues, one has to refer to the selective history again, by which the process has *become* knowledge. So, in the end, one returns to selection to get a satisfactory explanation.

Skinner (and many of his followers) held that this selective approach reaches far beyond simple behaviors, such as a pigeon's pecking a lever to get food (Skinner, 1953). The principle of selection-by-consequence does, in their view, not only apply to activities that immediately lead to beneficial consequences but also to "preparatory" activities that lead to activities that lead to beneficial consequences. Moreover, this scheme can be reapplied any number of times: Preparations can be for preparations. If this is true, a Skinnerian approach might help account for activities that exhibit a complex teleological structure: activities where one is doing X in order to do Y in order to do Z, in order to do . . .

Skinner also claimed that his approach did not disregard-the "inner" aspects of mental life. He accepts persons might think, or feel pain, without overt movement. Here behavior seems absent, while mentality is present. Skinner's response to the worry that his system can't deal with such events is twofold. On the one hand, he claims that silent thinking, or feeling pain are just behavior too. It just occurs without movement. Secondly, he holds that if private behavior is ever to be explained, it has to be explained in the same ways that outer behavior is explained. If we want to explain why someone thinks a thought privately, we have to invoke the circumstances in which the person thinks the thought, whereby the circumstances include the person's entire history (Skinner, 1953, chapter 17). Doing this, we might find out that the person keeps the thought hidden, because they think this holds an advantage, or avoids a sorry consequence. And then we can repeat the exercise for why they think this second thought, and so on.

According to the Skinnerian approach, anything that cannot be explained by a gradual story of selection has no place in scientific psychology. Skinner rejects that a capacity for thought could emerge as an autonomous trait through a genetic mutation in organisms that previously lacked it (Skinner, 1981, p. 503). This rejection conforms to the overall evolutionary framework of Skinner's times, which assumes graduality.

A Skinnerian selective story of a complex capacity would require showing how the capacity could derive in a gradualist way from more basic capacities (which themselves can be shown to derive from yet more basic capacities, and so on). As an example, consider what Skinner calls "precurrent behavior." A person shows "precurrent behavior" (Skinner, 1953, 1966) when, in attempting to solve a problem, they produce changes in the environment that will make it more likely that they will arrive at a solution. They might rearrange objects, jot down lines of thought, or seek an environment that has proved to be productive in the past, perhaps a busy café. Such precurrent behavior, analogous to exploratory activity observed in many animals, might have been selected as a particular operant activity (Mook et al., 1993; Neuringer, 2002; Page & Neuringer, 1985). Precurrent behavior in humans often involves self-stimulation using language, as when one formulates the steps one is going to follow to reach a certain goal and then follows the procedure spelled out.

Clearly, Skinner believed language use could be explained by his selection-by-consequences approach. He devoted a book-length treatment to it, in his work *Verbal Behavior* (Skinner, 1957). Skinner emphasized that language is special, in that it is essentially social. That is, the reinforcement processes that are involved in the emergence and most functioning of language inherently concern multiple persons. Without language, it is the consequences of one's own actions that reinforce or punish the actor. Reinforcement in language often involves the actions of other persons. One asks someone to pass the salt, and, if the request is followed, one obtains the reinforcing effect because of the actions of another person. Skinner defined verbal behavior as all behaviors reinforced through other persons. It isn't restricted to speaking or writing. A knock on the door that makes someone let you in counts as verbal behavior too.

Skinner stresses that language emerges in a social matrix in which reinforcement is widely distributed. Reinforcement can come in varying ways. The issuer of commands can obtain reinforcement by having commands obeyed, the responder by hearing "thank you," "I'm very much obliged to you," by a cooperative attitude, or by avoidance of punishment in return. Skinner puts commands together with demands and requests in a category he calls "mands." "Tacts" function differently, assuring contact with events might not have witnessed oneself. They function as signs for objects or events and, in human language, evolve into elaborate descriptions of situations. They have potential reinforcing value because they allow hearers to adapt to situations that they have not themselves perceived or partaken in. You might "tact" me that there are no trains today, and I might start looking for an alternative. I might return the favor later by informing you of something useful to you that you wouldn't have found out by yourself. By keeping the practice of tacting alive, members of a community sustain a system with great mutual advantages for the community.

Skinner points out that one of the major benefits of language is that it allows for the formulation of descriptions of rules for acting. Language makes it possible for us to learn how to act appropriately by listening to descriptions. This spares us the burden of having to find out by ourselves how to act accurately—if we ever would find out, we might not succeed or die in exploring. With language, people can teach other people where to find shelter, how to prepare raw foods so that they become digestable, fight effectively with a weapon, build a cathedral, or protect against malware. Because of language, says Skinner, much of human behavior is rule-following rather than contingency-shaped (Skinner, 1966, and, somewhat remarkably, Place, 1988). Contingency-shaped behavior is behavior that has been learned through direct experience, for example, through a process of trying and adapting when one doesn't succeed—where one obtains no reinforcement or gets punished. Language is the means par excellence to make possible and sustain rule-following behavior. Language allows lessons drawn from contingency-shaped behavior to be passed along from those with direct experience to those without. It allows for the stabilization and accumulation of knowledge beyond individuals and beyond generations, which is a defining feature of human culture (Skinner, 1981).

4.6 Against Behaviorism

Though Burrhus Skinner was a professor at Harvard until his retirement in 1974, and was the most influential psychologist for decades, he was the target of criticisms coming from various directions. There has been a tradition of criticizing behaviorism ever since it appeared. Such tradition grew strong and became dominant with the rise of cognitivism in the second half of the twentieth century (see chapter 7). This growing dissatisfaction with behaviorism in psychology, led to applying the label "behaviorism" indistinctively to a variety of approaches, from Watson to Skinner. Such behaviorism was associated with a number of claims, such as:

- all actions, of animals and humans alike, can be analyzed as specific reactions to specific stimuli;
- only ontogeny matters for psychology, not phylogeny;
- all learning is associative as in Pavlovian conditioning;
- learning only occurs when by blind trial and error a solution is found by chance;
- the mind can never be inner or private;
- thinking, interpreting, and imagining do not exist.

These complaints can be coupled to empirical work. One classical source can be found in the work of Wolfgang Köhler (1887–1967). In 1921, Köhler published a book based on his experiments on problem-solving in apes (Köhler, 1921). For example, he let chimpanzees enter a room with food hung from a rope descending from the ceiling. The banana was too high to reach by grasping, even by jumping. But the room contained boxes, and after a while, chimpanzees started to put the boxes on top of each other until they were high enough so that they could be climbed to grasp the food. Köhler analyzed this phenomenon in terms of the acquisition of "insight" and claimed this factor could not be accounted for by trial-and-error learning.

Other work that became very influential in the critical assessment of behaviorism is to be found in the work of Edward Tolman (1886–1959). Tolman was himself an experimental and theoretical psychologist who operated in the behaviorist tradition. Some of his work concerned rat navigation in mazes. Tolman described, in early (Tolman & Honzik, 1930a,b) and later (Tolman et al., 1946) experiments what he called "latent learning." Latent learning should be seen in contrast to learning that is the result of direct reinforcement. For example, Tolman describes that rats that have run around in a maze in which only one path leads to food seem to have learned the structure of the maze, even if they have received food only after a number of trials without food. So, it seems that they learn even when no food is given, and the lessons learned in the foodless trials come into play after they later receive food (Tolman & Honzik, 1930a,b).

In further work, and connecting it with his later (Tolman et al., 1946), but not his earlier (Tolman & Honzik, 1930a,b) experimental work on maze learning, Tolman speculated about the need to invoke the existence of "cognitive maps." Tolman writes:

> In the course of learning something like a field map of the environment gets established in the rat's brain. . . . The rat in running a maze is exposed to stimuli and is finally led as a result of these stimuli to the responses which actually occur. . . . The incoming impulses are usually worked over and elaborated in the central control room into a tentative, cognitive-like map of the environment. And it is this tentative map, indicating routes and paths and environmental relationships, which finally determines what responses, if any, the animal will finally release. (Tolman 1948, p. 198)

Further, Tolman thought there were not only cognitive maps for spatial navigation, but in a move that he described as "cavalier and dogmatic" (Tolman, 1948, p. 207), he proposed to assume the existence of a much wider array of cognitive maps, operative for example for navigating one's social environment. At the end of his paper, he writes:

> Over and over again men are blinded by too violent motivations and too intense frustrations into blind and unintelligent and in the end desperately dangerous hates of outsiders. And the expression of these their displaced hates ranges all the way from discrimination against minorities to world conflagrations.
>
> What in the name of Heaven and Psychology can we do about it? My only answer is to preach again the virtues of reason—of, that is, broad cognitive maps. (Tolman, 1948, p. 208)

There are many concerns that can be raised about the broad antibehaviorist movement that emerged, both as general criticism of behaviorism, as well as a criticism of Skinner. In the first place, it treats very different psychologies and philosophies of psychology as the same. Criticisms that apply to one might not apply to the others. It can and has been argued that criticism referring to phenomena such as insight learning or latent learning cannot serve to disqualify specific forms of behaviorism such as Skinner's approach, because this approach *does* have resources to deal with the relevant observations (Jensen, 2006). For example, Skinner's approach can refer to long-stretching processes of selection by consequences in phylogeny and ontogeny to explain why "insight" or "latent learning" occurs. Earlier selection by consequences, instead of reinforcement during the experiment, might have made rats prone to avoid certain places or structures in mazes in experiments such as Tolman's. Strategic choosing of pathways might have evolved as a general strategy in animals because it was, overall, beneficial. Similarly, tendencies for explorative behavior and perception with an eye to further effects might have evolved during phylogeny and ontogeny, and this might form the basis of insight learning (see, e.g., Skinner [1984] on problem solving). In other words, someone defending the Skinnerian approach can argue that the cumulative nature of selective evolution in general spills over to selection by consequences of organismic activity. Finally, a theorist invoking "insight" or "cognitive maps" saddles themself with the plight to account for where "insight" or "cognitive maps" come from and how insight processes and cognitive mapping work. For example, if a certain structure in the brain is identified as a cognitive map, how does it map whatever it maps? How do other processes in the brain use this map? Do these processes "read" and "understand" the map and apply this understanding? If so, what is this "reading" and "understanding" such that neural processes—instead of persons or organisms—can do it? If no convincing answers are provided to such questions, entities or processes such as "insight" and "cognitive maps" play the role of unexplained explainers. Or, in Daniel Dennett's term, a "cognitive loan" has been taken, that should at some point be paid back (Dennett, 1971).

There can be little doubt that the most consequential piece of criticism of behaviorism and Skinner, is the review linguist Noam Chomsky published of *Verbal Behavior* in 1959 (Chomsky, 1959). It is widely reprinted and described in many subsequent texts as offering "devastating criticism"—very frequently with exactly that phrase—not just of his theory of language but of Skinner's entire approach and of putting behaviorism out of the game.

Skinner himself never responded to Chomsky. After reading the beginning of Chomsky's text, he put it aside as beside the point. Other people have sprung to Skinner's defense. A detailed analysis and rebuttal of Chomsky's critique was offered by Kenneth MacCorquodale (1969, 1970).

It is instructive to look at some of the main points of Chomsky's critique, while also paying attention to MacCorquodale's response. One of the targets of Chomsky's critique concerns the term "stimulus." Chomsky criticizes Skinner for not using the term in an objective way. Instead, so Chomsky claims, Skinner's broad use of "stimulus" allows him to smuggle in assumptions belonging to what Skinner elsewhere rejects as "mentalism." For example, Chomsky argues that if the response "Eisenhower" can be made to both the man himself, as well as a thought about him, or as a result of imagining him, the notion of "stimulus" has lost the kind of objectivity that Chomsky thinks Skinner's behaviorism is committed to. In fact, Chomsky claims that some of the stimuli have been "driven back into the organism," turning Skinner into a closet "mentalist." MacCorquodale points out Skinner is explicit that the same response can be made to a variety of stimuli. The only condition for the response "Eisenhower" being explainable in Skinner's system is that there is a history of reinforcement that makes such response probable in particular circumstances. Such history could include having learned to respond "Eisenhower" by imagining him, or by "seeing" him in his absence. Here how Skinner dealt with "private events" becomes relevant (Skinner, 1953, chapter 17). Skinner claims his approach is still able to explain responding by first imaging or seeing-in-absence if a selection-by-consequences story can be given of imagining or seeing-in-absence. That is, imagining might be dependent on the circumstances of the imagining, and by identifying these circumstances, Skinner would try to explain why someone first imagined "Eisenhower," and then verbally responded by saying "Eisenhower." So, if Skinner would "drive the stimulus back into the organism," he immediately "drives it back out." He contends that, if the second move—the driving out—is not made, one is left with an unexplained explainer: imagining or seeing-in-absence that superficially appears to explain the overt response but actually explains nothing because it is itself left unexplained (Skinner, 1953, p. 278).

More generally, MacCorquodale criticizes Chomsky for falsely attributing a "one-stimulus-one-response" view to Skinner, on which Skinner couldn't account for the fact that one might respond to the same stimulus in different ways. However, Skinner has many ways of recognizing the role of context in explaining a current activity. Activities are dependent upon the whole history of an organism's interaction with its environment, and situations contain many aspects that the organism might engage with. Operants come to form a complex interrelated web or network, and many episodes of operant conditioning exert mutual influences.

Chomsky criticizes Skinner for not allowing that the activity of speaking must be explained by invoking preexisting organismic structure. In particular, Chomsky claims that speaking grammatically can be explained only by assuming that the brains of speakers are prewired to produce and understand speech in certain ways. Chomsky's claim is systemic, in the sense that it is part of a wider connected set of assumptions about how cognition in general and language and grammar in particular should be studied (more discussion of these will be offered in chapter 7). Yet the general point underlying the critique repeats the concern that Skinner's behaviorism necessarily overlooks what in a response is due to properties of the organism, instead of properties of the current environment the response occurs in.

As MacCorquodale points out, the Skinnerian reply to such concerns is basically always the same. Either such properties, for example a preexisting inclination with respect to speaking, is explainable via a selective history, on the phylogenetic (species) or ontogenetic (individual) timescale, or it is not. If it is, it is within the grasp of the selection-by-consequences approach. If it is not, one has introduced as alleged explanatory factor, something that itself requires at least as much explanation as it delivers. Nothing has been gained. Skinner could admit, if there were evidence for such, that organisms have a neurophysiologically based predisposition for language production and understanding. However, such predisposition would have to be explainable as the result of selection by consequences—its consequence being acquiring language (MacCorquodale, 1970, p. 96).

MacCorquodale accuses Chomsky of operating with a reified notion of grammar. Without mentioning Dewey, MacCorquodale argues Chomsky commits the kind of fallacy that Dewey warned for in "The Reflex Arc Concept in Psychology" (Dewey, 1896). That is, in both speaking and understanding activities one can discern a sensitivity to the grammatical aspects of language. But it is one thing to construct grammaticality as an aspect of speaking and understanding and another to assume that it is some "thing which the child, and later the adult, has and uses." (MacCorquodale, 1970, p. 94, also p. 96). The second assumption that in Chomsky's work takes the form of holding that the child has "an underlying theory of grammar that underlies both speaking and understanding" doesn't follow from the first step. MacCorquodale offers an analogy. It is not because a child learns both to walk and to recognize walking in others that both capacities should be explained by a common causal factor: "a theory of walking which he uses in his perceptions and in his activities" (MacCorquodale, 1970, p. 95).

Again, Chomsky's critique is systemic. It hangs together with general ideas about the mind known as "cognitivism" (which will form the topic of chapter 7). Cognitivists typically emphasize the importance of "inner mental states" for understanding the mental and hold that these are unexplainable by behaviorism. Chomsky couples a general cognitivist attitude to very specific views on a capacity for grammar as preexisting actual language use both in phylogeny and ontogeny (see chapter 7 for more).

4.7 Evolution Evolving

Despite Chomsky's specific commitments, there is a general line running through his critique that resonates with how evolutionary theorizing has been and is developing. The concern is that Skinner's overall approach and, by consequence, his treatment of language is too schematic and lacks specificity. To see this, consider that Skinner makes explicit that both in his selection-by-consequences approach and in his account of language, he aims to explain and describe psychological and linguistic phenomena with as few concepts as possible. In the context of language, Skinner's reductionism shows up in the fact that he self-consciously commits himself to explain language in terms of the notions that he uses in his account of behavior in general, such as operant and reinforcement (Skinner, 1953, chapter 1). What is special about language, according to Skinner, is the particular way in which reinforcement occurs: not through one's own actions, but through the actions of other actors.

But is it wise to wish for a strictly limited vocabulary when it comes to explaining action, mind and language? Can all the work be done by only a handful of concepts? To see how this concern applies to Skinner's evolutionary ideas about mind and language, we have to take a step back to evolution in general. As said, the modern synthesis has long been the standard view. But it has come under increasing pressure from new developments in evolutionary theorizing. Many contemporary evolutionary theorists think the modern synthesis is overly reductive. It paints a schematic picture of evolution in terms of a limited set of concepts: of gradual adaptation, due to random genetic change, to pressure from the environment. More concepts are needed to understand evolution than are available in the modern synthesis. Interestingly, contemporary theorists pay attention to the evolution of mind and language. They argue that a broadly Darwinian understanding can be given of mind and language. But at the same time, they point out that evolutionary accounts of mind and language need specially tailored conceptual tools. Though their overall approach to mind and language is not unlike Skinner's, it calls for conceptual expansion.

The case that the modern synthesis should be seriously revised is made by Kevin Lala and coauthors in their book, *Evolution Evolving* (Lala et al., 2024). It connects to earlier evolutionary theorists such as Conrad Waddington, Susan Oyama, Richard Lewontin, Mary-Jane West-Eberhard, Eva Jablonka, and Marion Lamb, and relies on large body of experimental evidence and theoretical analysis (Jablonka & Lamb 2005; Lewontin, 2000; Oyama, 1985; Waddington, 1957; West-Eberhard, 2003). An overall line of concern is that the modern synthesis is too abstract and general. It is too much focused on traits-in-general related to an unspecified environment coded by abstract genes that are susceptible to random change-in-general. Evolution becomes an "algorithm" that gradually filters out organisms so that only those adapted to the environment remain.

By adopting such a generalizing attitude, many of the specific details of actual biological evolution are overlooked, so the authors of *Evolution Evolving* argue. Real evolution and selection happens because concrete organisms lead life cycles and deal with environments in many different and specific ways. The concrete properties of organisms and their environments cannot be always abstracted from when one wants to understand the dynamics of evolution. Evolution occurs to something that has structure and properties of its own. Just as various sculpting material—clay, wood, or marble—do for an artist, the material nature of what undergoes evolution offers both resistance and opportunities. Adaptation is determined by what undergoes adaptation. Whether and how a trait evolves depends on what the trait is and what organism it is a trait of. Traits hang together. In the developmental process that leads from a single cell to a mature organism, several traits often develop together: They are produced by the same process and changes in the process lead to changes in multiple properties at once. Instead of intervening on individual traits one by one, evolution can take larger steps because, with a change in a developmental pathway, a number of traits change together. For example, the precursors of blind cave fish seem to have evolved quickly from seeing to adapted to a life in lightless cages. Several traits in those precursors were affected by one or few mutations, but the result was a fish having several properties adapted to the new cave environment (Lala et al., 2024, chapter 8). Some traits have become connected during evolution—for example, by being part of a common developmental pathway. They have changed and will change together. A result of this is that developmental pathways can, in later stages of evolution, be "retaken of the shelf" and previous structure can reappear when circumstances change again (Lala et al., 2024, p. 177).

Lala and colleagues want to replace gene-centrism with a focus on organismic development, in a broad sense. Development includes not only physical growth but also activity of organisms in natural and social environments. Genes play an important role in these developmental processes, but the metaphor of the gene as a blueprint or a complete code for the organism or its development falls short. Though genes play roles in developmental processes, they are one among many other factors. How a single cell grows into a complex organism is determined by a multitude of causes, including many nongenetic ones. Whether and how a particular gene influences development can depend, for example, on ambient temperature—as does the sex of a crocodile's embryo—or the presence of certain chemical substances in the organism or its environment. The same genes give different results in different circumstances.

Inheritance too is not a matter of genes alone, according to the view of evolution presented and argued for by Lala and colleagues (Lala et al., 2024, chapter 10). What matters for inheritance is that the next generation shares properties with the previous generation. Preservation of organismic structure and features can result from many other processes besides the passing on of genes (Lala et al., 2024, chapter 10). Organisms intergenerationally transmit nongenetic biological entities, such as gut-dwelling microorganisms, and these contribute critically to host physiology and function.

Organisms often grow up in the same environmental structure as where the previous generation lived. Such environmental structures are often made by the organisms themselves—a process called niche construction (Odling-Smee, 2024). The burrows made by Darwin's earthworms offer an example. A structured environment provides newly arriving organisms with action possibilities that they don't have to create by themselves. Burrows contain processed soil that can be used, and the new generation of earthworms doesn't have to invent burrowing or processing the soil. Niche construction and its inheritance is widespread. Inheriting a constructed niche provides human drivers with the possibility to easily get from one city to another by just following road signs (see chapter 10). Niche construction shows that in evolution, organisms are not passively molded by their environment but actively shape the environment, which then again shapes them. Organisms are both objects and subjects of evolution. Organismic activity is both a result and a cause of evolution (Lala et al., 2024, p. 240).

A process of inheritance that is of crucial importance to understand human evolution is social or cultural learning (Lala et al., 2024; Boyd & Richerson, 1985; Henrich, 2016; Laland, 2017). Through it holds a critical place in human lives, it is found in various animal species (Lala et al., 2024, p. 250, note 28). For example, separate groups of humpback whales have developed different ways of collective hunting for schools of fish, involving thumping with their tails on the surface of the water in one group, and creating "bubble nets" in the other group. Different groups of humpback whales show different behavior because they—perhaps originally one individual—have hit upon different strategies, which are transmitted within and between generations of the group (Lala et al., 2024, pp. 4–5).

Cultural learning relies on individual learning. Referring to Skinner's work, and very much in line with the selection-by-consequences approach, Lala and colleagues note that organisms have capacities for acquiring learned behaviors, which "is often the result of an exploratory search conducted over multiple trials, with selective retention of rewarded outcomes, a process that resembles natural selection" (Lala et al., 2024, p. 282). Cultural or social learning allows for retaining and transmitting the results of the trials of many individuals and makes them available "for free" to a larger group. Lala and colleagues point out that imitation or copying is not one generic thing but comes in different shapes. For example, red-winged blackbirds copy feeding strategies of conspecifics, but not when the conspecific shows an aversive reaction to the food (Lala et al., 2024, p. 116).

In their treatment of language, Lala and colleagues emphasize, like Skinner, that language offers a superior and unique tool for retaining and transmitting in time and space what is learned (Lala et al., 2024, chapter 13). Copying or imitating only goes so far; with language, the capacity for storing acquired knowledge becomes almost boundless. However, in the account of language promoted by Lala and colleagues, specifics matter very much (see also Henrich, 2016; Laland, 2017). Language evolved in humans due to a combination of many specific factors, pressures, and opportunities. These include anatomical changes in the vocal tract, circumstances favoring cooperation among individuals, ways of sharing visual attention on the same focus, sensitivity to when actions conform to a pattern, or deciding who to trust as teacher (Henrich, 2016, chapter 13). Many of these aspects rely on specifically human bodily properties or specific ways of acting. For example, the process of joint attention by a learner and a peer is thought to play an important role in language learning. It has been pointed out that the conspicuous white part of the human eye—the sclera—might have evolved to facilitate tracking where another person visually attends to (Tomasello et al., 2007). To give another example, a clever experiment of Horner and Whiten (2005) revealed that children but not chimpanzees copy irrelevant actions, such as ticking a box, before taking out a reward, a phenomenon that has been labeled "overimitation" (Hoehl et al., 2019; both experiments are reported in Lala et al., 2024, p. 325). This shows that imitation learning in humans is specific, relying on placing great trust in a teacher (see also Henrich, 2016, chapter 13). Moreover, the specific form that human cultural evolution has taken leads to specific genetic changes. Genes adapt to culture too. A process of gene–culture coevolution occurs (Lala et al., 2024, p. 220).

Overall, these contemporary approaches raise concerns about behaviorism and Skinner's theory by exposing its schematic nature and its relatively limited stock of explanatory concepts. They criticize modern-synthesis-based approaches to evolution as too abstract and operating with a meagre understanding of evolution. More is at play than blind mutation responding in a piecemeal fashion to selection pressure from the environment. Evolution has produced organismic structures that play a productive role in evolution. Evolution evolves and during evolution ways of evolving have appeared that did not exist 3.7 billion years ago. An analogous criticism applies to Skinner's evolution-based account of mind. Though contemporary evolutionary thinking about cognition, culture, and language recaptures several Skinnerian themes and ideas, it does so in a framework that is much enriched compared to the one that Skinner adopted.

But it is a different question whether these and other criticisms form a sound basis for the complete rejection of "behaviorism," including Skinner's approach, that has become commonplace since roughly halfway the twentieth century. As Louise Barrett has written: "Behaviorism Is Not Satanism" (Barrett, 2012, 2016). In the case of Skinner's behaviorism, if one looks through the mist of suspicion that has been cast upon it, one can find an elaborate attempt to apply the evolutionary thinking of his times to psychology. Something that might also be noticed when a more clear-eyed view of Skinner is aimed for is that his whole approach is motivated by practical goals. For him, psychology does not end with the achievement of a body of knowledge but with the achievement of means to improve the lives of individuals, societies, and humanity as a whole.

4.8 Evolution in Philosophy: John Dewey Again

Evolution was originally applied in biology and then taken up for psychology and cultural development. But some theorists have argued that the consequences of evolutionary thinking have profound implications for philosophy. John Dewey, the American pragmatist who we encountered in the previous chapter, wrote profusely on this idea (Dewey, 1910).

In particular, he thought evolutionary thinking uprooted a traditionally dominant way in which philosophy understands itself. In this traditional conception, philosophy is concerned not with truths but with what truth is, not with our theories of reality but with reality itself, not with existing moral rules but with what is objectively good. Philosophy concerns what is necessary, objective, and certain; it aims for the general, not for what is concrete. Thus, there is a division of labor between philosophy and the sciences, and philosophy is supposed to be able to attain, in its domain, what the sciences cannot attain.

Ancient Greek thought offered grounds for supposing that thinking could reach behind the concrete phenomena. There was supposed to be an objective order, factual and moral, pervading reality. Rational beings could, through their capacity for reasoning, get access to that order. Even for an empirically-minded philosopher such as Aristotle, both the physical and biological world were driven by the inherent general nature of things. Heavy things have a heavy nature and therefore strive to find their natural place close to the heaviest of objects: They fall to the ground, if given the chance. Seeds develop into plants, trees and organisms to realize their inherent nature.

On Dewey's account, the scientific revolution of the sixteenth and seventeenth centuries removed the idea of reality as driven by inherent natures from physics, but only with Charles Darwin's theory of evolution, this picture of reality came under pressure in biology. But removing inherent forces from science is one thing, removing them from philosophy another. By distinguishing philosophy from science, it is still possible to cling to a view of philosophy as able to reach valid conclusions about the objective metaphysical and moral order of reality. However, so Dewey held, Darwin's lessons for biology should be taken seriously in philosophy too. Philosophy should look into a Darwinian mirror and revise the image it has of itself.

To do that, philosophy should rethink its picture of reality. Dewey argues for prioritizing the concrete over the abstract. Concrete things exist; abstract concepts are ways in which the concrete can be categorized—the existence of abstractions is secondary with respect to the existence of what is concrete. In fact, Dewey diagnoses the tendency of metaphysicians to prioritize what is abstract and fixed over what is concrete and changeable in the same way he diagnosed it, in his famous "Reflex Arc" paper, in thinking about perception (Dewey, 1896). With respect to perception, Dewey warned for construing sensations, which in reality are only aspects of concrete perceptual processes conceived of in thought, into concrete components or, even worse, causes of perception. In his plea for Darwinian thinking in philosophy, he warns for a similar mistake, writ large: assuming that abstract aspects of specific things—those aspects that we can identify by comparing and analyzing specific things—are the existential or metaphysical grounds of these things. After Darwin, metaphysical modesty should be paired with epistemological modesty. Philosophers can't afford to treat what experience and experimental work shows as if it doesn't meet some higher standard for knowledge that is the privilege of philosophers.

Behind Dewey's plea stands a conception of reality as never fixed, open, and always in the making. Just as species of organisms evolve, reality as a whole is, according to Dewey, an open-ended project. And so is philosophy.

It seems that this radical historicity is the deepest philosophical lesson Dewey thinks can and should be drawn from Darwin.

Bibliographical Note for chapter 4

William Baum's *Understanding Behaviorism. Behavior, Culture, Evolution* offers a wide and sympathetic introduction to Skinnerian behaviorism (Baum, 2005). Louise Barrett, Mecca Chiesa, and Vicky Lee all try to go against the widespread intellectual unease about Skinner (Barrett, 2012, 2016; Chiesa, 1994; Lee, 1988).

Edward Reed's *Encountering the World: Toward an Ecological Psychology* pays considerable attention to Darwin's earthworms (Reed, 1996).

I have not mentioned the line of thinking that self-advertises as "evolutionary psychology," with as main proponents John Tooby and Leda Cosmides. Though prominent at the end of the twentieth century, it is by many judged now to have been too uncritically committed to a narrow set of assumptions about cognition and evolution. Darcia Narvaez and colleagues have argued that the kinds of development in evolutionary theorizing that were offered at the end of this chapter show the limitations of "evolutionary psychology" (Narvaez et al., 2022).

Note that Kevin Lala and Kevin Laland are one and the same person, who changed his name back to that of his ancestors.

5 Unity Through Translation and Identification: Logical Postivism and the Identity Theory

> When we run over libraries, persuaded of these principles, what havoc must we make? If we take in our hand any volume of divinity or school metaphysics, for instance, let us ask, Does it contain any abstract reasoning concerning quantity or number? No. Does it contain any experimental reasoning concerning matter of fact and existence? No. Commit it then to the flames, for it can contain nothing but sophistry and illusion.
>
> —David Hume (1748/1999, section XII, Part III, p. 123)

5.1 Unity, with Divergence

Consider again the athlete, of whom we say they have recovered physically but not mentally. We've seen in chapter 2 how dualism capitalized on the difference between the physical and the mental. In Descartes's language, it concerned a "real distinction," one reflecting not only a separation one can make in thought but a distinction between things that can exist separately. But many thinkers about the mind have taken an opposite stance. They argued that the difference between the physical and the mental is not absolute. It is not a difference between fundamentally different kinds of things, or fundamentally different aspects of reality, but a difference between what is at some level of analysis the same. Reality, at its core, is one. On such a view, everything that happens—from atoms whirling around to a plant growing to you reading these words—happens in one and the same space and time. There is a profound unity to everything that exists.

One might find these formulations vague. Indeed, as we'll see, some philosophers of "logical positivist" persuasion have attempted to clarify the concept of unity, without resorting to broad, sweeping statements about "reality" and its "fundamental nature." They have attempted to obtain unity in language instead of in "reality." They have argued that minds are not special—or not fundamentally special—because what can be said about minds can be *translated* in a language that does not contain seemingly problematic mental terms. When we talk about an athlete being physically in shape, we talk about processes and conditions of their muscles, lungs, blood, and so on. When we talk about them being mentally in shape, we talk about processes in and conditions of their brain and about how we expect them to act—not exercising properly or not sticking to the proper diet.

Other people have pursued the unification program in a different way. They have not shied away from making statements to the effect that the mental and the physical are just one. They are known as mind/brain identity theorists. Their central claim is the opposite of Descartes's thesis of real distinction: The identity theorists hold that mind and brain are nondistinct. Even if in thought we can separate mind and brain, in reality they are identical.

The views of the logical positivists and the identity theorists on the mind remain relevant for at least two reasons. First, unity remains at the core of our experience. As Descartes realized very well, most of the time, we don't experience our minds as separated from our bodies—our minds are not like sailors on the ship of the body. Minds develop as bodies grow and lose strength when the body weakens. We intervene on minds by exercise, medication, or surgery. We practically deal with mind and body as unified, not as disconnected. Second, both identity theorists and, more radically, the positivists question the validity of philosophical problems or established ways of philosophizing about the mind. They might help us to realize that certain philosophical problems—such as the supposed inexplicability of consciousness—may not be absolute but rather contingent upon numerous additional assumptions, some of which may be mistaken.

5.2 Logical Positivism and Psychology as Part of a Unified Science

The origin of logical positivism (or logical empiricism) lies in the meetings of the so-called Vienna Circle (*Wiener Kreis* in German) in Vienna at the beginning of the twentieth century. The main figures of the Vienna Circle were Moritz Schlick (1882–1938), Otto Neurath (1882–1945), Hans Reichenbach (1891–1953), and especially Rudolf Carnap (1891–1971). In Great Britain,

A. J. Ayer (1910–1989) was the disseminator of this philosophy (in his early publications, in particular *Language, Truth and Logic,* 1936). A well-known later logical empiricist is Carl Hempel (1905–1997). The logical positivists were influenced by Ludwig Wittgenstein's *Tractatus Logico Philosophicus*—though Wittgenstein (1889–1951) considered himself misunderstood by them (Wittgenstein, 1922).

A major motivation for the logical positivists lay in their awe for science as one of the only domains and methods in which it was possible for humans to reach agreement. Their view of science was based on theories on linguistic meaning and on knowledge. Building on earlier philosophies going back to David Hume, Immanuel Kant, and many other thinkers in the Western tradition, they distinguished between synthetic and analytic statements. Synthetic statements are factual; analytic statements are conceptual or definitional. To grasp the difference, contrast "The door is red" with "Red is a color." When one says, "The door is red," one describes something. Whether the sentence is true or false depends on whether the door actually is red—whether that fact holds in reality. Synthetic knowledge is a posteriori. One needs to look at a particular door to know about its actual color.

In contrast, when I say, "Red is a color," I state something about the relations of words in a language with respect to each other. "Red is a color" is true, but it is not true because of a fact about the door. It is true because of the ways we use the words "color" and "red." It concerns the rules of language, or how we describe reality and not how reality happens to be. Knowing how to speak English implies that one knows that red is a color. Knowledge of a language gives one a priori knowledge of its analytic truths.

"The door is red" is a synthetic statement. It says something about reality that goes beyond what one knows when one knows the meaning of its terms. For all the rules of language, the door could have been green, blue, or yellow. "Red is a color" is analytic, because it is already true because of the meanings of the concepts "color" and "red." "Red is a color" is always true, independent of what actual colors things have. No observation of reality is needed to assess whether it is true or false. The meaning of "color" contains, so to say, the meaning "red" (and that of all other colors), and therefore the statement "Red is a color" is "analytic." Just by analyzing meaning, by tracking the rules of language, one can find out whether an analytic sentence like that is true.

These views on meaning and knowledge lead to a position on what can be meaningfully said. According to the logical empiricists, what can be meaningfully said either is synthetic and concerns factual knowledge or is analytic and concerns the conventions of language. Scientific statements are meaningful because they indeed are analytic or synthetic. Logic and mathematics are purely analytic. Logic and mathematics consist of defining concepts and tracing out the consequences of such definitions. If one defines a circle as all the points at the same distance from a given point, one can easily see that any point closer to the given point than the stipulated distance must lie within the circle. In logic one defines the word "and" by pointing out that two statements joined by the word "and" are true if both of the constituting sentences are true, and false otherwise. From this, it is easy to figure out that if I say "It is Tuesday and it rains," this would be false if it rained, but it was a Wednesday. Most sciences have, besides relying on logic and mathematics, a factual or empirical component. Observation and experiment deliver synthetic statements, for example, about the melting point of a physical substance or about the three-dimensional shape that a certain molecule has.

One can now see why logical positivists placed their trust in science as a source for agreement among humans. Either one can find out about the truth of synthetic statements by observation and experiment, or one can find out about the truth of analytical statements by logical, mathematical, or definitional analysis. Everyone should agree on what is observed, or seen in front of their eyes, or on what is true by definition.

What is the place of philosophy in all of this? Logical empiricists did not regard philosophy as a science, but rather as a discipline or practice aimed at thinking about knowledge and along the way reform science so that it can play a beneficial role in society. On the logical empiricist agenda, the aims and methods of philosophy should be primarily oriented toward the advancement of knowledge. This means that philosophy should abandon several traditional projects and should set up or continue other ones. For the logical empiricist, the philosopher's role is to promote clarity, to rigorously analyze language—particularly scientific discourse—and to prevent errors in reasoning that stem from linguistic misuse or unfounded assumptions about language.

The logical empiricists thought philosophers had been and were particularly prone to make mistakes that could be revealed by an adequate understanding of meaning and knowledge. They targeted in particular *metaphysics*. The logical empiricists saw metaphysics as the branch of philosophy that aimed at revealing deeper or more fundamental truths about reality than those that science could deliver. Metaphysicians, on this view, are concerned with such questions as "What is the nature of reality?," "Is reality at bottom one, or are there fundamentally different kinds of reality?"—one senses the affinity with the opposition between monism and dualism here.

From the logical empiricist standpoint, these metaphysical ambitions are doomed from the start. There is no deeper insight into reality than what can be gained by science. If there was, it would be, by definition, unscientific. When people, philosophers in particular, think there are deep metaphysical questions, they are, so the logical empiricists contended, deluded by language. They use words in ways that are unusual and mistaken. As a result, questions get formulated, and problems get raised that cannot be solved, not because the answer is too difficult to find but because the question simply makes no sense (Carnap, 1931). Unanswerable questions can result from confusing the modes of language. One could mistakenly think that a factual, synthetic statement can be proven a priori. Or one could mistake an analytic statement for a synthetic one—as when one would go on a hunt to find evidence for "Red is a color" by observing doors. Or one can mix up questions about the role of terms in language for questions about reality. According to logical empiricism, such would happen if one would ask "What is being?"—a question central to the work of Martin Heidegger (see Heidegger, 1927).

In Carnap's view, here one makes the mistake of assuming that the word "being" refers to a property in the way that "red" or "yellow" do. If one makes this assumption, the question "What is being (or existence)?" would be equivalent to the question "What is red?" According to Carnap, arguing in the footsteps of Kant (1781), such equivalence does not hold, because, in its normal, correct use, the word "being" does not function as a predicate. A predicate is a word that refers to a property. The predicate "colored" refers to all colors; the word "red" refers to the property red. "The door is red" consists of a word for a thing, namely the door, referred to by "door" and the predicate "red" referring to the property red. The word "is" acts to bind predicates to subjects and thereby ascribe properties to subjects or things, as happens when one says, "The door is red." So "is," a form of "to be," has a particular role in language, which is different from the role of predicates. It can be shown that assuming that "is" can be treated like a predicate is a mistake by pointing out that a sentence like "The door is is," or "The door is being" is nonsense, in contrast to "The door is red." But that would not be so if "is" acted as a predicate—then "is" would be a predicate and "The door is is" would make sense in the way "The door is red" does. Because "red" is a predicate, one can meaningfully pose the question, "What is red?" The answer can be provided by science, by physics or psychophysics. But because "is" (and other forms of "to be") have a different use from that of predicates, the similarly shaped question "What is being?" only seemingly makes sense. No answer at all can be provided to it, apart from pointing out that the very formulation of the question rests upon confusions about language.

So logical empiricists thought that it was in science, not metaphysics, that one would find what's true about reality. Moreover, a further tenet of logical empiricism, strongly promoted by Carnap, was that there should be, at bottom, only one language to express all the truths of all sciences. Carnap and other logical empiricists strived for a "unified science," pertaining to the whole of reality. In such a unified science, sentences in what we know as physics, chemistry, and biology would all be expressed in the same language. Logical empiricists called the unified language "physicalist," because it was meant to be a language for describing facts, as exemplified by the science of physics. With the unifying physicalist language, the facts of psychology would then be expressed in the same physicalistic language in which the facts about physics, chemistry, and biology were formulated.

Carnap sketches how such renderings of psychological sentences in a physicalistic language would go (Carnap, 1933/1959). He considered as an example of a psychological sentence: "Mr. A is excitable." He pointed out that the sentence is analogous to a sentence in physics like "Y is firm." Carnap regards both statements as *dispositional.* They invoke a tendency something has to show a certain pattern of responses or behaviors in certain circumstances. Carnap thinks "firm," applied to material objects is a dispositional term, invoking the tendency of firm objects to resist deformation when hit by something, for example. Moreover, he thinks that firm objects show this tendency because they have a particular physical structure—a specific way the atoms from which the object consists are connected to each other. The term "firm," then, is dispositional and refers to an underlying structural property to which certain tendencies are due.

Carnap thinks the same analysis applies to psychological terms, such as "excitable." Therefore, on Carnap's view, the psychological sentence P1 "Mr. A is excitable" is equivalent to a physicalistic

> sentence P2 which asserts the existence of a physical structure characterized by the disposition to react in a specific manner to specific stimuli. In our example,

> P2 asserts the existence of that physical structure (micro-structure) . . . of Mr. A's body (especially of his central nervous system) that is characterized by a high pulse and rate of breathing, which, on the application of certain stimuli, may even be made higher, by vehement and factually unsatisfactory answers to questions, by the occurrence of agitated movements on the application of certain stimuli, etc. (Carnap, 1933/1959, p. 172)

Carnap thus proposes to reformulate sentences containing mental terms in sentences containing neurophysiological and behavioral terms. In this way, the mental gets accounted for in a science-respecting way. This is how logical empiricism offers a program to deal with the mind. If the translation project would be carried out, nothing "mental" that is problematically different from what is talked about in the rest of science would remain. Problems about the place of mind in reality would have disappeared.

Two things should be noted here, because they will be of importance in the light of later criticism of Carnap (see chapter 8). First, Carnap's physicalistic counterpart of the psychological sentence refers to both the neural system and to movements and activities. Second, when Carnap says that the psychological sentence and P2 have the same meaning or content, he has a particular concept of meaning or content in mind. He is talking about what is known as "reference" instead of what is known as "sense" (the distinction goes back to Frege [1892/1948]). Reference has to do with what's the case in reality, considered apart from the way it is described. The descriptions "The writer of Romeo and Julia" and "The writer often called 'the Bard'" both have the same reference, because they are about the same person. Yet the descriptions have a different sense, in that they describe the same referent in a different way. One could understand one description, but not the other, because of one's knowledge of English.

Sense has to do with how concepts relate to each other in a language. Roughly said, sense belongs to the "analytical sphere." Reference is factual: If statements concern the same fact, they are referentially the same. Roughly said, reference belongs to the "synthetic sphere." What's important in the current context is to acknowledge the difference between reference and sense. To say that expressions in a language have the same reference does not imply that one says that they have the same sense. Carnap's claim that translations can be found for sentences containing mental language is a claim about reference. It is meant as the claim that a sentence not containing mental terms can be found that has the same reference as the sentence containing the mental terms.

The aim of Carnap's program or ideal of translation was not to provide an analysis of the sense of mental terms in languages like English. Carnap was aware that a dispositional analysis would provide only the reference of a term like "excitable," not its sense. There was no claim whatsoever, on Carnap's behalf, that speakers of English, by competently using the term "excitable," would also know its reference. Such speakers knew the sense of the term "excitable" but not necessarily its reference. This is not unique to mental terms: Speakers of English can say that a physical object is firm, without having any knowledge about which structural properties of objects make firm objects have the dispositional properties that make them firm.

The explicit comparison Carnap makes between "excitable" in psychology and "firm" in physics allows to highlight some further aspects of his analysis of psychological terms—again they will prove important in light of later criticism of the position ascribed to Carnap (Crawford, 2013, 2014). Carnap has always been explicitly aware that a dispositional analysis in terms of an underlying structure, such as applying to firm and excitable was *defeasible*. That is, it is always possible to find circumstances in which the conditions listed in the analysis occur with respect to the subject of analysis, without the analyzed property occurring. Or: exceptions are always possible, in special circumstances. Indeed, dispositional analyses are "incomplete" and "conditional" (Crawford, 2014, p. 15).

The thesis, held by Carnap, and also by Hempel (1935/1949), about "psychology in a physicalistic language" (Carnap, 1933/1959) is about language. Crucially, it should not be mistaken for the metaphysical position "that everything that exists is physical," or "that everything that exists is material." These further theses are, in Carnap's view, objectionable metaphysical credos. In line with his general criticism of metaphysics, Carnap didn't think they were, as they stood, acceptable. Once again, Carnap's case is based on an analysis of language, in particular on a distinction that Carnap makes between the *material* and the *formal* use of language (Carnap, 1934/1937). One uses language in the formal mode when one talks in language about language. To say that sentence P1 is translatable into sentence P2, that sentence P1 implies sentence P2, or that "red" is a predicate is to speak in the formal mode. Now it is possible to, mistakenly, phrase something that should be explicitly in the formal mode without presenting it explicitly in that mode. Then one says something that actually is *about the language* used to describe facts and things, but on the face of it, looks like it is *about facts or things*. In such cases, statements seem factual, but they are not. An example of a statement that looks like a statement about things, while in reality it is a statement about language is: "Having blue eyes is a property" (Cirera, 1993, p. 4). This use of language is characterized by Carnap as being in "the material mode." To interpret sentences such as in the example "having blue eyes is a property" as being about things that exist in the world is to sow confusion. It can lead to attempts to give an account of these assumed objects, facts, or processes. For example, one is tempted to construe a theory about the nature of properties. This is to engage in metaphysics: One has been led to assume problematic existents, and now one has to forcibly find a place for them among the more familiar furniture of the world. But, on Carnap's view, such enterprise will lead nowhere because it is premised on the mistake of not having noticed that talk about properties is about language and consequently should be explicitly in the formal mode. The metaphysical questions about the problematic existence of properties need no answer. They can be avoided and replaced by solvable formal questions about language, its constituents, and its rules. Once a question such as "What is a property?" is formulated in the formal mode, it is transformed into a tangible problem. For example, "Having blue eyes is a property" can be rephrased in terms of the role property words play in language, that of predicates, terms that can be coupled to symbols for objects.

The distinction between formal and material modes applies to the relation between minds and bodies. If one asks how the mind relates to the body, one operates in the material mode and seems to be asking something about two different objects: minds and bodies. This immediately raises questions about what those objects are. Metaphysics comes knocking. If one opens the door for it, one might be further tempted to say things as "All mental events are identical to physical events." But this claim, and related ones, are rejected by Carnap as a bad, metaphysical version of "Sentences about mental events are translatable into sentences about physical events." So, the lures of metaphysics can be resisted, on Carnap's playbook, by transforming statements and questions about the relation between minds and bodies to the formal mode. They then can be seen as being about the relations between sentences that contain mental terms and sentences that do not contain mental terms.

5.3 From Identification to Identity

The textual origins of the mind/brain identity theory are easy to locate, namely in two papers that were published by Ullin Place (1924–2000) and Jack Smart (1920–2012) (Place, 1956; Smart, 1959). Smart himself presents his paper as an elaboration of Place's. The core idea of the identity theory sounds straightforward enough: Sensations are brain processes. That is, the experience of pain, or the sensation of red, are identical to the occurrence of a brain process. In its original form, the identity theorists restricted their aims to sensations. They did not claim that, for example, thoughts were identical to brain processes. In fact, they were sympathetic toward (a certain understanding of) the views of the philosophers that will be treated in the next chapter. Smart later revised his position and thought the identity approach applied to more or all mental phenomena. Place never did; he expressed support for scientific behavorism, in particular, Skinner's version, when it came to everything mental beyond sensation (see, e.g., Place, 1988).

Perhaps the idea that sensations are brain processes might look rather meager or trite to form the backbone of a philosophical theory. But there is philosophical substance to the work of the mind/brain identity theorists. The identity theorists were very much concerned not just with stating the identity claim but with defending it against philosophical a priori arguments that say it *cannot* hold. Two kinds of a priori arguments were targeted by them. In the first place came arguments to the effect that there is some logical error involved in claiming that conscious experiences are identical to brain processes. One such alleged error would consist in ascribing properties to conscious experiences that they by definition cannot have. By stating identities between mental phenomena and brain processes, one would formulate absurdities. In this vein, one could argue, against the identity theory, that the feeling of hurt when one cuts one's finger is in the finger and, therefore, cannot be in the brain. Or one can hold that a feeling of melancholy is not located at all, while according to the identity theory, it should be located at some place in the brain. But this is absurd, for one thing cannot at the same time be not located and located at a specific place. In short, mental concepts and terms for brain processes have different kinds of meanings, they obey a different logic, and, therefore, they are incompatible. The assimilation at the very core of the identity theory is simply but plainly impossible.

Besides targeting these logical arguments, the identity theorists aimed at a line of reasoning that relied on the nature of introspection, of the way we are aware when we experience. Experience, so this reasoning goes, shows itself to the experiencer *as* mental. What subjects are confronted with in experience are mental inner objects or properties. And one can go on, arguing that these mental properties cannot be reconciled, let alone identified, with physical properties because we know the physical only by being acquainted with mental properties. On this line of reason, just the fact that we experience already disqualifies the idea that mental phenomena could be brain processes.

The heart of the identity theory consists in attempting to neutralize these two types of argument. The case for the identity theory made by Place and Smart rests on the idea that what we know at some point in history about the mental might be incomplete or wrong. People might think certain things about the mind, but it might turn out that what they think is not true. The identity theorists point at science as a better guide to truth than what people think now. If science tells us that mental phenomena are located in space, they are located in space, even if people used to believe they were not.

5.3.1 The Logical Case

To get his project of proposing and defending the identity theory going, Place (1956) made a distinction between different senses of "is" in sentences such as "A is B." First, he made a gross division between the "is of identity" and the "is of predication." The "is of identity" is used to provide an "adequate characterization" of something. So, one can say that a square is a rectangle with four equal sides. Or one can say that a cloud is a suspension of small water droplets. These examples both use the "is of identity," but in a different way. In the first example, the terms joined by "is" have the same meaning, where meaning is understood as the "sense" discussed in the previous section. They are, to again use the terms introduced in the previous section, analytic. Therefore, to deny such an "is" is to say something contradictory. Given the definition of "square" as "a rectangle with four equal sides," one cannot say "This square has five sides." One would then at the same time say that a square has five sides (in one's statement) and four sides (in one's definition).

The second example, "A cloud is a suspension of small water droplets," is different in not concerning meaning (or sense) but composition. It says that clouds are composed of suspended droplets of water. It states a factual truth, not one about meanings/senses. It is, to use the vocabulary of the previous section, synthetic, not analytic. Both kinds of identification statements give adequate characterizations but in a different way. The characterizations are adequate in the sense that they are complete in a way that is absent in what Place calls the "is of predication," as when one says that Socrates is Greek. By saying that Socrates is Greek, one does not characterize Socrates in the same comprehensive way as one does when one says that a cloud is a suspension of small water droplets. In the "is of identity" one can, says Place, always add "and nothing else," but one cannot do so when the is of predication is at play.

The identity theorists then replied to the logical line of arguments against the identification of consciousness and brain processes that such identification was of the compositional kind. Because these identifications concern facts, not senses, one of the terms of the identification can have implications, for example, concerning localization, that the other term has not. Because the identifications in the mind/brain case are of the compositional kind, one can negate them without literally contradicting oneself. That is, even if consciousness is in fact a brain process, one can have a language in which it is possible to deny without absurdity that it is one, just as one can deny that a cloud is a collection of suspended water particles.

Finding out the truth about the composition of things might require empirical investigation. In the words of the previous section, it is synthetic and a posteriori. For example, we now know that a strike of lightning consists of an electrical discharge, but this knowledge about what constitutes lightning could only be arrived at after the natural sciences had advanced to a certain level. It did not just step on the scene just by talking about lightning. Rather, people first talked about lightning, then made investigations into electricity, and only after a long history connected lightning to electric discharges. Even if lightning is an electric discharge, seeing lightning *as* an electric discharge required hard scientific work. Before the current knowledge was established, people also spoke of lightning, but without knowing anything about electrical discharges. Lucretius, the Roman philosopher writing in the first century BCE, claimed in his *De Rerum Natura* that lighting was composed of fiery sun atoms that winds pressed out of the clouds that contained them. Speaking about lightning therefore does not commit to claiming anything about electrical discharges.

5.3.2 Science and Explanation

Why, then, are philosophers so prone to mistakenly conclude that sensations *cannot be* brain states? The identity theorists see a crucial role here for different ways in which we can know about sensations. We can know about sensations through scientific observation, as when a neuroscientist observes processes in the brain, but we can also know about them by employing introspection. Importantly, we have no way to know sensations in those two ways simultaneously (Place, 1956). But it is a mistake to conclude from a difference in ways of knowing to a difference in what is known. Place writes:

> It is clear from this that the terms "cloud" and "mass of tiny particles in suspension" mean quite different things. Yet we do not conclude from this that there must be two things, the mass of particles in suspension and the cloud. The reason for this, I suggest, is that although the characteristics of being a cloud and being a mass of tiny particles in suspension are invariably associated, we never make the observations necessary to verify the statement "that is a cloud" and those necessary to verify the statement "this is a mass of tiny particles in suspension" at one and the same time. We can observe the micro-structure of a cloud only when we are enveloped by it, a condition which effectively prevents us from observing those characteristics which from a distance lead us to describe it as a cloud. Indeed, so disparate are these two experiences that we use different words to describe them. That which is a cloud when we observe it from a distance becomes a fog or mist when we are enveloped by it. (Place, 1956, pp. 46–47)

Still, Place notices that in the cloud/droplets example, the two ways of observing it—as a cloud and as a collection of droplets—might be less opposed to each other than the ways of observing are in the sensation/brain process case. It is true that, normally, when we see the cloud, we don't see the droplets. Conversely, when we see the droplets, we don't see the clouds. But in some cases, one might enter what initially looks from a distance as a cloud, and, once immersed in it, observe the water droplets. There might be a continuous path that could be followed so that the two ways of observing flow into each other. Such continuity in the process of observation encourages the identification of the cloud and the droplets. Perhaps we would be more prone to identify sensations and brain processes if such smooth means of switching from introspection to scientific brain observation existed. But given the absence of such, why should we identify sensations and brain processes?

Place, in addressing that question, returns to the lightning/discharge example. The reason we identify lightning and electrical discharge, so he says, is that we know that the occurrence of electrical discharge explains all observations of lightning. The electrical discharge affects our sensory systems in all the ways that the bolt of lightning does. Now, an analogous reason for the identification of a brain process and a sensation can be given: We can identify the brain process and the sensation if the occurrence of the brain process explains everything about the introspection of the sensation. The identity theory claims that this is the case: The occurrence of the brain processes affects our consciousness in the same way as does the occurrence of the sensation. Here we meet an argument for the identity theory that motivates it not by showing that reasons for rejecting the theory are wrong, but by providing a self-standing reason to accept it. The heart of this positive motivation is a claim about explanation: Just as the facts about lightning explain the facts about observations of lightning, facts about the brain explain facts about experiences, in particular, what happens when we introspect sensations.

But this crucial claim on behalf of the identity theory quickly leads Place to having to dismantle another objection. This objection is based on exactly the denial that brain processes explain introspection. The objector claims that what introspection deals with, namely experience, is a special *phenomenological* property, which is neither identical to a physical property of the environment, nor identical to a property in the brain. Phenomenological is meant here as "inherently and evidently not physical." The objector, as pictured by Place, illustrates the argument by the example of experiencing a green afterimage. Afterimages occur when one stares long enough at a colored light source. They tend to be of the complementary color of the light source, so a red light will lead to a green afterimage. Suppose you have looked at a red, roundly shaped light source. If you look away, you might experience a roundly shaped greenish afterimage. But where is the experienced green? The experienced color is not in the brain, so the phenomenological argument goes, because, as neurophysiological investigations might show, nothing in the brain is literally green when we experience the afterimage. But neither is the green in the objects of the environment, for what was seen was red. Instead of being in the physical world, the experienced green is a "phenomenal property" in the subject's "phenomenal field" (Place, 1956, p. 59), so the argument concludes. Of course, this conclusion shows only that the green is mental if being a phenomenal property implies being a mental property—but Place seems to regard this assumption as self-evident.

Place's answer consists in returning to the explanatory argument for the identity theory. Also in the case of the green afterimage, Place contends, the mind/brain identity can explain introspection. It can do so by linking the experience of the green afterimage to the experience of a green object in standard circumstances, that is, when an actual green object is present. Under those conditions, the green is in the outside world, or so Place says. If the brain process that occurs when experiencing a green afterimage is same or similar to the brain process that occurs in standard situations of seeing green, the identity theorist can explain introspection of afterimages too, so Place claims. Afterimages are just a matter of brain processes that happen outside of normal circumstances. Because they occur standardly when the sensed property is present, they are introspected in terms of that property.

But introspection is not blind to the fact that the situation is not normal. People reporting experiences of properties that are not present in their environment often qualify their judgments by saying, "It seems green to me," "It looks green" or "I experience green," instead of "It is green" or "I see green" (Place, 1956, p. 50). That is, though the same or a similar brain process occurs, people also notice that it now occurs in unusual circumstances. For example, it doesn't escape them that the afterimage moves with the movements of the head in a way that the objects in their environment don't. Or they close their eyes and still experience the afterimage, while they no longer see their surroundings. So, people, when introspecting in unusual circumstances, not only discriminate a particular brain process—the one correlated with green—but also have learned to discriminate the unusual circumstances from normal circumstances. Once this is spelled out, it becomes clear, so Place claims, that referring to nonphysical phenomenal properties as they can occur during the experience of afterimages is not valid. Everything that needs to be explained with respect to what introspection delivers in such cases, can be provided by the identity theory.

Smart (1959) follows Place's trail. He too gives the example of an afterimage, be it a yellowish-orange one. Like Place, he analyzes the experience of the afterimage in terms of the similarity of the experience in the afterimage condition, that is, when no yellowish-orange object is present, to an experience in a standard condition in which an object is present. Smart writes: "My suggestion is as follows. When a person says, 'I see a yellowish-orange after-image,' he is saying something like this: 'There is something going on which is like what is going on when I have my eyes open, am awake, and there is an orange illuminated in good light in front of me, that is, when I really see an orange'" (Smart, 1959, p. 149).

That way, the external object retains a primary role. The case of the "purely mental" experience is shown to be dependent upon a case that is not "purely mental" but involves worldly objects. The strategy pursued here by Place and Smart, is, in its overall contours, similar to the strategy used by B. F. Skinner to deal with privacy: allowing for some special category of mental events but construing them as dependent upon a more basic kind of mental events that are supposedly less problematic. Smart adds something, however. He makes the additional claim that when one speaks of one's sensations, one does not thereby express some commitment to nonphysical phenomenological facts. Introspecting one's sensations is, Smart says, "topic neutral"; it is neutral with respect to the question of whether sensations are thought to be inherently nonphysical or, on the contrary, physical.

5.4 Two Steps Further: Eliminative Materialism, Twice

Thinking about the identity theory, one might be struck by a tension within it. On the one hand, the identity theorists want to assert that, at the bottom, everything is physical, so that, strictly speaking, there exists no distinct mental realm, nor any exceptional mental properties or entities. Yet, at the same time they claim that mental processes are identical to physical processes. But to qualify for identification with something else, that which is identified—the mental process—has to be real. If something—such as a mental process—does not exist, it cannot be identified with anything. Moreover, if something—again, a mental process—is identified with something real, this seems to affirm the existence of what is identified with the something real, in opposition to its denial. Paul Feyerabend, widely known for his "anarchistic" position in the philosophy of science, took note of this, describing it as "the dilemma of any identity hypothesis concerning mental events and brain processes" (Feyerabend, 1963, p. 295). Again, the dilemma is that by mentioning the mental in any identification sentence, one accepts the existence of the mental. Moreover, identities go in two directions. So, if one asserts that the mental is physical, one both physicalizes the mental and mentalizes the physical. Physicalizing the mental is, of course, the aim of the identity theorists, but mentalizing the physical is the opposite.

We are running into exactly the kind of problems that Carnap wanted to avoid with his distinction between formal and material modes of speech. In Carnap's view, questions such as whether mental processes are identical to physical processes are impossible-to-solve metaphysical problems. They are impossible to solve not because we lack knowledge but because we have formulated a confused question. Once we switch to the formal mode, the question changes its nature, and the intractable aspects disappear. We then no longer ask how the mental process of pain can be identical to a brain process, but we ask how we can translate sentences in which the word "pain" occurs into sentences that contain only physicalistic vocabulary, such as neurophysiological and behavioral terms. Arguably, Smart attempted to neutralize the issue Feyerabend put his finger on by proposing that our standard use of words such as pain is "topic neutral," or not committed to assuming the existence of a separate mental realm.

These possible answers by Carnap and Smart to Feyerabend's "dilemma" did not move him. Feyerabend's line is that, if one wants to dispel the mental from reality, one also has to dispense with the mental in language. Instead of continuing to use the word "pain" on one side of statements about translatability or identity, we should simply stop talking about pain—and the same goes for other mental terms. We should forget our mentalistic vocabulary and replace it with nonmentalistic vocabulary and theory, just as we ditched talk about "being possessed by the devil" in favor of talk and theory about "epilepsy" (Feyerabend, 1963).

If Feyerabend's proposal would be followed, mental vocabulary would be eliminated from the scene and replaced by neurophysiological vocabulary. Instead of reporting that we are in pain, we would say that a certain brain process was occurring. Eliminativism was further developed, in a rather nuanced way, by Richard Rorty (1965).

Eliminativism was later revived by Paul and Patricia Churchland (Churchland, 1981, 1986). They motivated eliminativism by attacking the alleged theoretical role mental terms play, in particular, expressions such as "believing that" or "desiring that." Their eliminative materialism depends upon an understanding of mental talk as theoretical, of beliefs, desires, and intentions as alleged "causal internal states," and on other assumptions that have become articulated and prominent only well after Feyerabend published his short paper on the topic. Being dependent upon conceptions of the mental and its vocabulary that will be treated only in further chapters, discussion of the eliminativism of the Churchlands will be postponed until chapter 8.

5.5 Subsequent Criticisms and a Shared Assumption About Language

All the theories and ideas presented in this chapter have come in for a lot of criticism. Indeed, there are often-repeated, well-established lines of rejection for them. Often, however, the target of the rejection is something quite different than what is taken for the targeted theory. In parallel, much attention is seldom devoted to what motivated the thinkers discussed previously. In chapter 8, we'll see that Carnap and Hempel are often considered to be "logical behaviorists," and therefore some standard and simple lines of criticism suffice to put their views aside. But "logical behaviorism," as it is defined in these rebuttals, is not the view held by Carnap and Hempel. Crucially, the "logical behaviorism" on which the standard criticism is based holds that mental concepts and descriptions can be translated without loss of sense, or are analytically equivalent to behavioral terms of description. In addition, it is often assumed that physicalistic language only concerns movements and omits reference to neurophysiological processes. But Carnap, and Hempel did not hold that translations were sense-preserving and they did not exclude reference to conditions of the nervous system or body. The identity theory and eliminativism have come in for their own lines of rather stereotyped criticism. We will deal with that in chapter 8, for it requires setting up the specific framework for thinking which feeds the criticisms.

In chapters 6, 10, and 11 we'll meet thinkers who take aim at a different aspect of the identity theory as proposed by Place and Smart. Place and Smart proclaim identities of mental phenomena and brain processes. Place restricts such mind/brain identity claims to sensations, but Smart includes much more on the mental side. The philosophers and theorists covered in chapters 6, 10, and 11 will question the centrality of the brain. They will oppose "neurocentrism" and argue that the locus of mental processes extends beyond the brain, the head, and even the body.

All the approaches to the mental discussed in this chapter, including eliminativism, despite their differences, share something. They all assume that mental terms are used to *describe* conditions of the body or brain. They hold a referential or descriptive view of mental vocabulary, a view according to which terms such as pain are mainly used for describing sensations or other mental phenomena. Though the referential view is deeply ingrained, it has not remained uncontested. In the next chapter, we'll turn to philosophers who target precisely this referential assumption about mental vocabulary, along with other deeply entrenched opinions concerning the mind. They are Ludwig Wittgenstein, Elizabeth Anscombe, and Gilbert Ryle.

Bibliographical Note for Chapter 5

The epigraph to this chapter shows how the overall lines of logical empiricism can be traced back to David Hume. His "reasoning concerning matters of fact" corresponds to synthetic statements, and his "reasoning concerning relations of ideas" corresponds to analytical statements. The affinity to Hume is made explicit in Alfred Ayer's *Language, Truth and Logic*, which provides a schematic and somewhat simplified tour of logical positivist thinking, written with the panache of a manifesto (Ayer, 1936). Logical empiricism as developed in the hands of Carnap, became very nuanced, elaborate, and often technical. A critical treatment of logical positivism, with attention to views of the mind, can be found in Hanfling (1981).

As will be discussed in chapter 8, with the rise of functionalism since the 1960s, the identity theory became a minority position. Still, it has been defended by Thomas Polger in his *Natural Minds* (Polger, 2004) and has inspired an approach to the sensory aspects of perception based on psychophysics by Austen Clark in his book *Sensory Qualities* (Clark, 1993).

6 Difference: Wittgenstein, Anscombe, and Ryle

I'll teach you differences.
—William Shakespeare (1623/2007, p. 40)

6.1 United for Difference

Now that we have seen two strands of thinking that, in their own ways, strived for unity in understanding the mind, three philosophers that warned for disregarding differences when it comes to thinking about the mental will be discussed. These are Ludwig Wittgenstein (1889–1951), Elizabeth Anscombe (1919–2001), and Gilbert Ryle (1900–1976). Though there are differences between the views of these philosophers, the broad outlines of their philosophies are quite similar. All three share similar views on philosophy, language, mental concepts and the phenomena involved in their use.

6.2 Bringing Philosophy Back to Life

"Look, don't think" is an often-cited advice Wittgenstein offered to philosophers (Wittgenstein, 1953, § 66). His aim is to urge philosophers to resist the temptation of hasty theorizing. When a philosopher begins to think about topics such as meaning, sensation, perception, or thought, they might look for what they consider safe ground. Convenient places to set off might be found in what is taken to be agreed upon by all, or something that "cannot reasonably be doubted." An example of such allegedly safe departure zone might be Descartes's certainty that he is thinking. Or a philosopher might look for a paradigmatic example of conscious experience and proceed by analyzing looking at a red patch. Perhaps the philosopher will then turn to introspection, and to the innards of their mind: what they are currently experiencing. And this might be the conscious feel of the red patch. Having isolated some presumedly uncontroversial truths or facts, or a paradigmatic token of the phenomenon of interest, the philosopher might build from there, neglecting, at least for the moment, the circumstances that normally or often surround the phenomenon. In fact, these might be seen as distracting from the core phenomenon, or as complications that can be brought in later. So, in analyzing language, the philosopher might end up analyzing pure meaning as the relation between a symbol and a thing in the world. Thought might be addressed as a pure mental occurrence, "That what stands before my mind now," and sensation can become the punctuated feel of pain or red, irrespective of the needle that causes the pain, the finger that the pain is felt in, the red raspberry seen, or the eye that is exploring it. After all, if the philosopher is interested in the feeling, why not just consider the pure feeling in itself? If considering the question of how the feeling relates to the brain, why not just consider the feeling and the parts of the brain that are active when the feeling occurs?

Wittgenstein and like-minded philosophers object to precisely these supposedly innocent preparatory steps, allegedly taken before one starts the real philosophizing. In their view, instead of clearing the ground for the serious work, unrepairable damage has already been done, when these moves are made, and the philosophy that is to follow will be inadequate. In Wittgenstein's words, "the decisive movement in the conjuring trick has been made" (Wittgenstein, 1953, § 308). One reason such philosophizing has started on the wrong foot is that it has assumed that the phenomenon of interest can be philosophically investigated without due consideration of its context. That would be true if the phenomenon contained some intrinsic property, which always occurred, perhaps with variations, in different contexts. But this needn't be true. Perhaps the phenomenon occurs only in certain circumstances. That is, even if we are able to single out the phenomenon in speaking or thinking, it might actually occur only when the full context is available. Consider money and its value (an analog Ryle repeatedly returns to; see, for example, Ryle, 1954, pp. 94–95). If one completely disregards the historical, social, or cultural circumstances in which money genuinely operates and just considers a $1,000 banknote, one might end up being perplexed by the question: How is it possible that this simple printed piece of paper has so much value? In fact, one can continue, why does it have any value at all? And why does *this* piece of paper—a monopoly note, or a photocopy of a bill—have no value at all? One might be led to assume a yet unexplained property of "valueness" or "pure value" that makes for the difference. Only, one has to find out how it does. By selecting a paradigmatic example and focusing their efforts solely on it, the philosopher has confined themself to a "one-sided diet" (Wittgenstein, 1953, § 593). Again, this move restricts where thinking can go, even before it has started to move. For perhaps there are other bona fide members of the class of things one is considering that are nevertheless different in many respects. Consider a patch of red. As it appears in philosophical thinking about color, it is supposed to belong to a surface one is facing—as when one is seeing a part of a red wall, or a wall with a red patch on it. But one can see color without there being a surface, as when one looks at a blue sky. One can be dazzled or blinded by a red light, but not by looking at a red patch. So how can one expect to acquire an understanding of red by just considering a situation in which one stands in front of a patch? Note that there is not one standard patch present whenever we speak of a patch. And even if it all worked out for red, why assume this understanding of red would lead to an understanding of color? Colors differ: Some colors are shiny, some get clearer with more light, while others like grays and blacks change or disappear when more light becomes available, and fluorescent colors seem to emit light. In these and many more ways, colors differ among themselves.

In fact, Wittgenstein famously invented the concept of a "family resemblance" to illustrate just how different things could be, while still belonging to the same category (Wittgenstein, 1953, § 65–71). He pointed to how the individual members of a family could look alike: Person A could have a similar appearance as person B, and B as C, and so on, but in the end, A and F might look very different. The likeness typical for the family does not consist in a core set of similar traits but of many interconnected local overlaps, in the way a thread hangs together, namely by fibers connecting across a small distance, not by one invisible inner thread that holds everything together from beginning to end. In the same way, there might be two activities called "games," without having anything in common.

Instead of isolating targets of philosophical attention such as experience, meaning or thought from the concrete circumstances in which they appear, the phenomena have to be considered in their natural contexts, so

Wittgenstein, Anscombe, and Ryle contend. To get a grip on experience or meaning, philosophers should look at animals or persons that are experiencing, or at how people talk to each other, act when talked to, make sense of what the other says, or teach a language to someone who does not yet master it. It is the active lives of organisms and the socioculturally shaped "forms of life" of persons that the philosopher must turn their attention to before they can expect to get an accurate view of experience, meaning, and many other phenomena (Wittgenstein, 1953, § 19).

It is in the midst of life that many of the phenomena that have raised philosophical curiosity and perplexity are to be found. If they are kept in their normal surroundings, the temptation to find these phenomena perplexing, expressing itself for example in raising philosophical doubt about their very existence, should disappear. Doubting itself, the topic of a later work by Wittgenstein titled *On Certainty*, is a special kind of activity that is only possible within a solidly structured sociocultural framework (Wittgenstein, 1969). "I doubt whether the table exists," on Wittgenstein's view, is comparable to such expression as "I declare you two married," or " I promise to pay off my loan from now on." These expressions make sense only in certain circumstances in which certain habits, customs, and institutions already exist. There has to be a practice for people to marry for two people to get married. There has to be money and practices of lending for it to be possible to express an intention to start paying off a loan. Without these surrounding conditions, the expressions make no sense. In Wittgenstein's view, the same goes for doubting. It might be perfectly meaningful to doubt whether a table exists in some circumstances. For example, one might have paid a sum for an antique piece of furniture online only to find out that it is not being delivered on the expected date, and that the seller is no longer reachable. But the concept of general and universal doubt—or doubt detached from any concrete circumstances, as Descartes envisaged, and applied to everything—is as problematic as the concept of a marriage or a debt per se.

If, in one's philosophical thinking, one disregards the situation and surrounding circumstances, one will lose track of important distinctions. If one considers only what unites things, one might lose track of differences that are in many contexts, for some purposes, more important than similarities. Consider, for example, visual perception and visualizing something in one's imagination. In Hume's philosophy of mind, and many subsequent treatments, seeing and imagining are thought of in terms of the occurrence of impressions and ideas—ideas being considered to form something like Hume's "faint copies" of impressions. In this treatment, despite the difference in vivacity, episodes of perception and imagination are taken to be of the same kind. But such a heaping together of perception and imagination misses a crucial dissimilarity that Wittgenstein points at: Perceivers have a different attitude to what they perceive than imaginers have to what they imagine (Wittgenstein, 1980b, § 85, see discussion and further references in Ter Hark, 2009, pp. 595–596). The perceptual attitude is observational. Perception is aimed at investigating what is independently there. One cannot steer perception in the way one can steer imagination. One can actively investigate what one is seeing and bring previously unseen parts into view by making movements, but this forms a means of uncovering what was there already before one moved. In imagining something, the attitude taken is different. Even if imagining needn't be initiated voluntarily—some visual image can pop up in one's mind—the imaginistic attitude is not one of observation. One cannot move to passively *find out* about how the imagined object looks from a different perspective, even if one can perhaps try to actively visualize what the result of moving when actually seeing the object would be. Further, one has a liberty to change what is imagined that one does not have with respect to what is observed. One may, by an act of will, alter the color of an imagined object, but one cannot by mere volition repaint a perceived physical object. A Humean analysis might provide a basis for skepticism—a position Hume himself was drawn to, even if he also argumentatively resisted it. If perception and imagination are fundamentally the same, how do I know that I am not imagining, or that my alleged perceptions are not impressions generated in me by an evil creature—a demon or a mad scientist? But once the distinct attitudes in perception and imagination that Wittgenstein draws attention to are not overlooked, the attraction of skeptical scenarios diminishes or vanishes.

Or, to take an example based on Anscombe (1957, § 2–3), consider the utterances, "It is going to rain" and "I am going to town tonight to visit an old friend." Superficially, they both look like predictions. One might want to refine that assessment by adding that in the case of the second sentence, the prediction concerns a fact in which oneself is an actor. Yet, as Anscombe points out, the differences between the sentences run deeper. When they turn out to be wrong, they do so for different reasons. When it does not rain, despite me having predicted it would, I might perhaps not have gathered sufficient evidence, or maybe the weather is just being quirky. That it didn't rain was not, however, something I could have done much about—even if I perhaps could have made a more accurate prediction. But when I say I am going to town to visit an old friend, and in the end, I stay home, what falsifies my utterance is not that I lacked evidence or that I am as unpredictable as the weather. Rather, by saying what I did, over and above predicting something, I expressed a commitment to a course of action. I might have later revised that commitment in light of further considerations, or I might have tried, but failed—perhaps there's a bus strike—but failing to live up to a commitment is not a failure of relating appropriately to evidence. Moreover, while the first utterance is factual, the second one is expressive, and by making it, the utterance becomes a factor in what I will do.

The focus of language use in Wittgenstein, Anscombe, and Ryle has led people to describe them as "ordinary language philosophers." But such a label is misleading. What these philosophers want to point to is the necessity of considering language use in the complex circumstances of forms of life, before one comes up with philosophical reflection on certain concepts or phenomena. Their program is not to provide an inventory of actual uses, or an attempt to freeze all possible uses to align them to current use. They always operate against a background of the philosophical denial of use. Wittgenstein coins the term "language games" for the various ways people put language to use (Wittgenstein, 1953, § 7). But he does so to advance philosophical insight, rather than to initiate empirical investigations of the specific games we currently engage in. The point is always that only practices of use in forms of life give sense. It is recognized that these practices change, differ, and could have been different. Wittgenstein, Anscombe, and Ryle often think about variations of our life-forms to draw conclusions about how different life-forms would give rise to differences in language use.

In the following sections we'll see how the attitude of "bringing back to life "(Mac Cumhaill & Wiseman, 2022), can be applied to language, sensation, intention, understanding, and the mind in general.

6.3 Difference in Language

Philosophical insensitivity to circumstances, and hence to difference, often has systemic origins. It is supported by a whole connected system of assumptions about a variety of subjects: general views of reality, language, thinking, feeling, science, objectivity, and many more. The resulting system of thought can be entered at various points. One entrance to decontextualized thinking about thinking is formed by a certain view of language—one that encourages abstractness and losing track of differences. This view at the same time motivates and is motivated by related views of, inter alia, reality and thinking. Because of the systemic nature of this view, one can adduce many reasons for it, and it can come to look compelling because of support provided from many directions.

Here is such a conception of language. It starts from the claim that meaning can exist only if there is a correspondence between language and the world. Hearing a sentence is hearing words, and understanding must consist in connecting the words and the world. Someone who understands a sentence must be able to construct meaning from the acoustics.

For such understanding to be possible, two conditions must be met. First, one must know what the words stand for. Second, one must know how to combine words to form meaningful sentences. Knowing the components is knowing what the individual words name or refer to. Knowing the rules is knowing which possible facts the combination of words in a sentence corresponds to.

To go from the word to the sentence level, there need to be rigid rules for constructing composite meanings from component words. I must master these rules to understand. But for the rules to exist and for me to master them, they must be objective. They can't be made together with language, but they must preexist for language to possibly have meaning. When you hear a sentence in your natural language that you have never heard before, you know what it means. But that can only be the case, so the reasoning goes, if the new meaning relies on a stable, already existing connection between language and world. If it were established together with the new sentence, you could never understand the new sentence. It would sound like a foreign language to you. Without objective rules, the correspondence between language and the world would get lost. Remember, complexes in language must correspond to complex possibilities in reality. If there was no objective combination of existents, the combination of words would lose contact with the world. It would be a string of words to which nothing discernible could correspond. It would fail to have meaning. So, there must be ways in which the elements of reality can combine that are mirrored in language. Mastering a language consists in mastering these rules.

On these lines of reasoning, the existence of this correspondence must be in place *before* we can use language. It is a precondition for the use of language. Without it, speaking and understanding a language would be impossible. The philosophical theory of meaning, or the theory about the general relation between reality and language that forms the basis of meaning, is entirely a priori. It need not invoke particular empirical facts about language use. The only empirical reality it needs to consider is that language, and the understanding of language, exists.

The core of these philosophical ideas about language is the general view of language and has been present in many strands of philosophical thinking. In the twentieth century, they played a role in the developments of logic, for example, in the hands of Bertrand Russell (1872–1970). The young Wittgenstein was fascinated by this view and, after contacting Russell, elaborated the view at great depth in his *Tractatus Logico-Philosophicus* (Wittgenstein, 1922).

Sometimes the view of language at issue, in general and as rendered in the *Tractatus*, is known as the "picture theory" or "picture-picture" of language. Wittgenstein (1922) himself used the analogy of language as a picture. The method used to arrive at the picture theory of the Tractatus was a priori. It started from what must be the case for language to have meaning. It offered an abstract theory of meaning that was meant to provide a foundation for the possibility of meaning. The *Tractatus* was not concerned with the specifics of actual language use: how people actually speak, in which contexts, and for which purposes. In this way, it did exactly what Wittgenstein later diagnosed as making a capital methodological error in philosophy. Notably, Wittgenstein added a dose of mysticism to his version of the picture theory. He claimed that the structural relations between the world and language underlying linguistic meaning could never be fully uncovered through language. He famously invoked the picture of a ladder that one stands on. In thinking about language, and its relation to the world, we rely on language and on how it relates to the world. Thinking can never allow for a totally detached view of what makes language meaningful. Because we are always in the midst of meaning, also when thinking philosophically about how meaning is possible, there remains an ungraspable aspect to meaning. We cannot throw away the ladder that we are standing on, without invoking a dose of mysticism.

In his later work, including the *Philosophical Investigations*, Wittgenstein offers a sustained criticism of the picture-picture of language and presents a different view (Wittgenstein 1953). He takes issue with the core idea leading to the picture theory: that the basic elements of language are words that are like names referring to elements of reality, and that sentences combine these elements so that the result pictures possible facts. Moreover, in such a view, these assumptions go hand in hand with the idea that the correspondence relation is abstract and necessary: It must exist for language to be possible at all. The meaning relation is more than something languages actually have. It is not something that comes into being with languages. On the contrary, the meaning relation must exist before there are languages.

Against this core idea of meaning as something objective, something before language, the post-*Tractatus* Wittgenstein insists that language is much more than modeling or picturing. Some language uses can indeed be seen as picturing possibilities, so Wittgenstein admits. One can say, "It rains today in Paris" and thereby picture a way that reality can be. Language can be a tool with which a rendering of reality can be painted. But not all language is of this kind. When I say, "From now on you are married," I do not picture a reality; I create one. When someone wins a prize and shouts out "Hurray," the person is not modeling something, but expressing relief or excitement in a situation. Many more examples could be (and are, in Wittgenstein's work) given.

In his later work, Wittgenstein stressed that standing in a pictorial relation to the world is not a precondition of language but something that language must achieve. When it is achieved, this happens after a long and complicated history. In other words, the capacity to use language as a descriptive medium is something that has emerged out of much more primitive forms of language use. Complicated rules for using language have gradually emerged out of simpler practices and activities. The kinds of highly complex languages that we now know have not jumped on the stage simply by clinging on to and exploiting some preexisting abstract correspondence relation. Instead, hominins that started uttering and reacting to uttered sounds had to live through a complicated history spanning thousands of generations until, from humble beginnings, our complex languages formed. Rules came about along the way. A path was made in walking, or better, in talking (van Dijk, 2016).

A basic assumption of the picture theory is that the elements of language—words—are like names for basic elements of reality. But the later Wittgenstein argues that even something seemingly so basic and simple as naming is extremely complicated. There is not one phenomenon "naming," nor is there an essence of "a name." There are names of persons, and nicknames, names of battles, names of historical time periods, and names of fashion trends (see Wittgenstein, 1953, § 36–37). These different kinds of names are embedded in different practices, and the practices determine what kind of name is at issue. One can use names to regulate a practice, as when a play's director says, "You are Blanche." Someone—coming from a different culture perhaps—could name persons but not understand what it would be to use a nickname or be clueless when they heard a name for a style of a piece of furniture. So even something seemingly simple and basic like naming is actually a very complicated something human beings do amid many other things done collectively. Naming can't be taken as the taken-for-granted fact on which a theory of language can then be built, as in the picture theory.

Hence, the idea that the fundamental workings of object words such as "triangle" and property words such as "green" can be readily grasped by assimilating them to naming is mistaken. Instead of being reducible to a basic and simple naming relation, using object and property words is massively complicated itself.

The complexity of using object and property words can be revealed by considering what it takes to learn such uses. Learning to understand or use object or property words requires a lot of stage setting: of already existing capacities and practices. The capacity to use such words needs to be acquired from a community. Repeating a word heard from a teacher in front of an object is just one of many things needed to learn an object word. And even doing that well requires a lot. You only know the word if you can produce it in the appropriate circumstances repeatedly. Learning sensory property words, such as color words, normally requires being confronted with objects that have the property, but much more besides that. Just by being shown a red object, a learner does not grasp the meaning of "red." Of course, one can have a natural, unlearned reaction to certain properties, but using a color word involves much more. One needs to be able to focus on the color only and disregard other aspects. One needs to learn that two surfaces with very different textures could both be called red—note here that a different culture could have different practices and build sensitivity to texture in its color vocabulary (for examples of such, see Gellatly, 1995). One must distinguish properties from objects—for example, the color red from the tomato—and be familiar with attributing different properties to the same individual, as well as the same property to different individuals.

Using property and object words acquires shape amid the many complexities of what Wittgenstein has called "a form of life." Some activities in a form of life do not involve language; others do. But language is interwoven and dependent on many of them, and once there is language, many practices change because of it. The very idea of a rule of language is only intelligible, in the Wittgensteinian view, in the context of such forms of life (Wittgenstein, 1953, § 19).

The assumption of abstract correspondence rules as a prerequisite for language, and as motivating the picture-picture, no longer makes sense to the later Wittgenstein. Such rules would float in the air; they are ghostly and mysterious. Such a priori, metaphysical rules make as little sense to the post-*Tractatus* Wittgenstein as would, for example, the idea that the rules of chess existed well before people started playing games. Such preexisting rules, either in language or in chess, don't explain anything because they are themselves unexplainable. Rules that grow out of practices, on the other hand, can be explained in a historical way that traces their development from simpler forms. What is needed, thinks Wittgenstein, to explain the rules of speaking or the rules of games, is not metaphysics but a natural history (Wittgenstein, 1953, § 415). A natural history tells how something originates from circumstances in which it was not yet present. Seeking explanations in a natural history is opposed to seeking explanations in a priori conditions that are assumed to be necessarily present before the event actually emerges.

Wittgenstein's *Philosophical Investigations*, his most famous book after the *Tractatus,* starts by demolishing the picture-picture of language and subsequently draws many conclusions for many philosophical questions. Often these relate to the philosophy of mind and psychology.

6.4 Difference in Mind

6.4.1 Wittgenstein on Sensation

Wittgenstein's thought itself is systemic: It consists of and relies on a lot of interconnected claims, examples, observations, and motivations. So, when Wittgenstein turns to the philosophy of psychology, his thinking connects in many and complex ways with his thoughts about philosophy, language, science and mathematics. Here, his approach to the mind will be shown by focusing on some strands in his thinking.

Let's consider sensation first. We've seen how sensations came to play a primary role in early psychology, a role that, despite a tradition of critics such as Thomas Reid, it has largely kept. We noticed how a specific concept of sensation entered psychology, building on Locke, Hume, and many other writers. Given the longstanding appeal of something like a picture theory of language, it has been tempting to many to think about sensations on what Wittgenstein calls the model of "object and designation" (Williams, 1983). Sensation words, like "pain" or "blue," are then thought of as names—or designators—for specific experiences—the designated objects. On this view, when I say I am in pain, I refer to something going on inside me, a pain or an experience of red.

But if one construes sensations on the "object and designation" model, one is bound to encounter complexities and problems. My pains and my color experiences seem in some respects quite unlike standard, external objects. When you and I designate the table we are both sitting at by the word "table," there is one object that we are both referring to. But when you say you feel pain or experience red, and when I say so, we are both using the same word "pain" or "red," but we are referring to different objects: You designate your pain, your red, I my pain, and my red. On a common construal in the case of sensations, the object referred to is internal and private. You have yours and I have mine. But then you know what you designate by "pain" and by "red," but strictly speaking not what I designate when I say the same words. The words are the same, but perhaps they refer to something different in you than in me. Locke capitalized on this possibility by invoking the thought experiment of spectrum inversion, or inverted color sensations (see chapter 3, section 3.2). Perhaps behind the same use of the words "red" and "green" reside inverted sensations in different people. Where the one experiences red, the other experiences green, while both use the word "red." Though the use of words is the same, the experiences are inverted. Nothing seems to stand in the way of such behavior being publicly well aligned, while the inner, experiential worlds are profoundly but systematically out of touch.

Even if a thought experiment like the inverted spectrum might seem far-fetched, the phenomenon it focuses on seems real enough. You don't experience my experiences, for sure. And both of us can hide experiences. I can say I felt or saw nothing, while I did. I might insist that I have no feelings or thoughts about revenge, but I might very well lie. I might keep episodes in my mental life, including sensations and experiences, hidden from you, and as long as I do, you might have no way of finding out about the truth. And you might do exactly the same.

So many seemingly innocent lines of thinking converge on a certain picture of sensations as inner, objectlike, and private. But should we accept it? We've already seen how the picture of sensations as inner objects was criticized by thinkers including Reid and Dewey. Wittgenstein's work contains elaborate lines of thinking and a sustained deconstruction of the alleged essential privacy of sensations, as well as of the idea that they should be thought of as inner objects. In his work, Wittgenstein questions the standard ways of picturing sensations as well as the motives for it, and he criticizes the support it finds in thinking about other subjects such as language. In short, Wittgenstein attempts to show that the seemingly undeniable picture of sensations as inner private objects leads to paradox, incoherence and inconsistency. That is, Wittgenstein thinks that the private inner object view inherits all the problems of the picture theory of language and brings intractable issues of its own. If the picture-picture doesn't work for language in general, all the more so it doesn't work for sensation language.

The sensation-as-private-inner-object view portrays learning to use sensation words as a process of naming objects. But now the objects are special, inner objects. Yet, Wittgenstein has already argued, in his reflections on language, that using words like "triangle" requires the existence of specific language practices, and a process of becoming skilled in the practice. Just being confronted with a triangular object and reacting to it doesn't suffice for mastering the use of "triangle." The process of getting the use right involves a temporally extended process of calibration, guided by language users that initially are more competent than oneself.

But in the case of the special kind of objects, namely, private internal ones, as sensations are supposed to be, additional problems arise. For in the case of external objects, as one learns the word "triangle," there are objects that can be identified and reidentified. But in the case of the sensation-object, understood as inner and essentially private, it is assumed that such stable factor does *not* exist. Your color experience could be the inverse of my color experience—your green experience, my red experience. Or it could be still something else. Indeed, as many philosophers have asserted, it could just as well be nothing. I might just be saying "red" in the right circumstances but have no sensation at all. All of this could be true if the sensation-as-private-inner-object view holds. But now, so Wittgenstein claims, the view has pulled the rug out from under itself. It tried to model inner on outer object word use, but it took away an essential element of outer object word use. Certainly, Wittgenstein recognizes that the outer object is necessary in outer word use. He doesn't think the object, when decontextualized and taken out of its surrounding form of life, *suffices* for such word use. But it certainly is necessary for the language game to get going. It just is one among other necessary factors. However, if one construes sensations as inner objects that could be anything, and even could be absent, one can no longer intelligibly work with the model of object and designation. The whole idea of sensations as inner private objects, construed on the object-designator model, evaporates (Williams, 1983).

Wittgenstein has illustrated this line of thinking with his famous "beetle in a box" thought experiment. It merits being cited at length:

> Now someone tells me that *he* knows what pain is only from his own case! Suppose everyone had a box with something in it: we call it a "beetle." No one can look into anyone else's box, and everyone says he knows what a beetle is only by looking at *his* beetle. Here it would be quite possible for everyone to have something different in his box. One might even imagine such a thing constantly changing. But suppose the word "beetle" had a use in these people's language? If so it would not be used as the name of a thing. The thing in the box has no place in the language-game at all; not even as a *something:* for the box might even be empty. No, one can "divide through" by the thing in the box; it cancels out, whatever it is.
>
> That is to say: if we construe the grammar of the expression of sensation on the model of "object and designation" the object drops out of consideration as irrelevant. (Wittgenstein, 1953, § 293)

But, if one has to leave behind the object-designator model for sensations, how does one have to think about them, according to Wittgenstein? The answer has already been given: by reconnecting them with the circumstances in which they occur. It is organisms that are hurt, that wriggle and shout out when hurt (Wittgenstein, 1953, § 284). It is only when we think we can isolate pain from the conditions under which it actually occurs that it becomes mysterious. Then we are led to questioning whether you too have the pains I have, or whether a stone doesn't have pain. But when we look at pain, instead of just think about it, these questions no longer make sense. Wittgenstein writes: "Look at a stone and imagine it having sensations. One says to oneself: How could one so much as get the idea of ascribing a *sensation* to a *thing*? One might as well ascribe it to a number!—And now look at a wriggling fly and at once these difficulties vanish and pain seems able to get a foothold here, where before everything was, so to speak, too smooth for it" (Wittgenstein, 1953, § 284).

In Wittgenstein's view, we use the word "sensation" to talk about aspects of the lives that organisms lead. But these aspects cannot exist outside of the lives of these organisms—just like "having a baroque look" cannot exist on its own, apart from concrete artworks or buildings. It is because they cannot stand on their own feet, so to say, that worries whether your sensations are different from mine, when our lives are sufficiently similar, don't make sense. For the sense of these questions depends on separating the sensation from the life, which is exactly what cannot be done in the Wittgensteinian view.

Further, aspects are not inner objects. To think of the pain of a person crying out in agony as an inner object that causes their movements is again to make the separation that Wittgenstein insists cannot be made without implying paradox and incoherence. It is the person, at this moment of their existence, because of what happened to them, and because it is a human, with a human life, involving all what human lives involve—human bodies, human desires, human activities, and human threats and liabilities—that is in pain. When human beings are in pain, they are in many ways like animals that are in pain, that is, insofar as there is overlap between the lives of humans and the lives of animals. But to the extent that human lives differ, there is room for difference among the pain sensations of humans and animals. Humans can talk about their pains, and this might lead to lend the object-designator model credence. But even in humans, so Wittgenstein points out, the use of language needn't be descriptive. People can cry out in pain, and when they do so, even when they use words such as "pain" or "hurts," their words do not play the role of designators of some (inner) object that is registered. Rather, the words are like an animal cry. They are ways to *express* the pain; they don't describe it. To express something is to do something, as when I exclaim "Yes!" in a burst of delight upon successfully accomplishing a task. Of course, humans can describe their pains, for example, in the doctor's office. But such is possible because humans have developed a more complicated pain-related form of life. The possibility of describing pain vindicates the designator-object model of pain only if one holds a picture theory of meaning.

But what about privacy? Is the private inner object view not correct in that respect? For can't one keep one's sensations hidden? Wittgenstein's response here consists not in denying that it is possible for a person, and perhaps other organisms too, to hide their feelings and experiences. Admitting that, the point he makes is that such hiding cannot be the rule. It must be the exception. Only what is normally not concealed can be hidden (see Ter Hark, 1990, chapter 5). The concept of hiding something requires that what is hidden could *not* be hidden, just like the concept of forged money requires the concept of real money (Ryle, 1954, p. 94). Just as it is logically impossible for only counterfeit money to exist, it is likewise impossible for there to be only hidden feelings or experiences. The object-designator model claims exactly such essential privacy for sensations. On the Wittgensteinian analysis, by doing so, it runs into another alley of paradox and logical confusion.

6.4.2 Anscombe on Intention

Elizabeth Anscombe (1919–2001) was a professor of philosophy at the University of Cambridge. Anscombe was interested in ethics. She was particularly unhappy with utilitarianism in ethics. Roughly speaking, utilitarians identify what's ethically good with what has maximally beneficial effects for the maximal number of persons or ethical subjects. In her view, this is despicable because it trades in absolute moral values for calculation. What is ethically bad can be declared good because it is beneficial to many. This is the way utilitarian thinking leads to justify the throwing of atomic bombs on Hiroshima and Nagasaki, and Anscombe strongly disagreed with it. However, she was convinced that serious ethical reflection required doing philosophy of psychology first. To ethically evaluate a person's actions, one must know whether the actions were intentional or not. But what is it to intend an action, or what is an intention?

With the most straightforward title possible, her book *Intention* is devoted to these questions (Anscombe, 1957). Like Wittgenstein, she wants to answer that question by "bringing philosophy back to life" (a phrase taken from Mac Cumhaill & Wiseman, 2022). That is, like Wittgenstein, she proceeds by attempting to get a grip on intention by looking at the lives of those to which the word "intending" applies: persons and animals. Leading the forms of life of animals and men, for example, by engaging in making "characteristically animal movements," or "movements with a normal role in the sensitive, and therefore appetitive, life of animals" (Anscombe, 1957, § 47, Mac Cumhaill & Wiseman, 2022, p. xiii), is a precondition for having intentions. Like Wittgenstein, she is convinced that the idea of disregarding the context, aiming for a "pure" treatment of "intention in itself" and related concepts, will bear no fruits and is bound to lead to perplexity and paradox.

She has illustrated this point with a discussion of the concept of "owing." She considers a situation in which a bag of potatoes is left at someone's doorstep (Anscombe, 1958). As a result, B, who is living at the house, owes the potato seller, A, a sum of money. But how is the fact of leaving the potatoes related to the owing? For that to become intelligible, one needs to know about practices of delivering goods to places of residence (nowadays quite different), which involve the deliverer being paid. But to understand that, one needs to know about economic practices and how money works. One also needs to know about mutual obligations, honesty, responsibility, and much, much more. In other words, only by seeing how the fact of leaving the bag at the doorstep is embedded in and interwoven with a wide-ranging complex of practices, one can see the connection between the fact and the owing. On the other hand, once the background is considered, once one understands the relevant practices which together constitute a "form of life," it becomes difficult not to see that the totality of facts *imply* the owing. "Owing" is a normative concept: It is about what one should do. Traditional philosophical wisdom has it that there is a deep gap between the normative and the factual. The realm of the factual concerns what is the case, and the realm of the normative concerns what should be the case. Many philosophers accept as an unquestionable principle what has become known as "Hume's cut," namely, that one cannot derive an "ought" from an "is." From the fact that monkshood is poisonous, one cannot conclude that it is bad per se, or that it is immoral to mix monkshood in someone's food. On Anscombe's contrary view, "Hume's cut" is not a fundamental philosophical principle but simply an artifact of philosophizing in a void. If one asks about "facts" and "norms" in the abstract, without placing them in the broad and variegated circumstances in which facts happen and norms are obliged, neglected, imposed, and rejected, one comes to such completely general conclusions as in "Hume's cut." But once one returns facts and norms to life, the connection between what is and what ought to be becomes totally transparent. One can confirm and explicate precisely why the inhabitant of the house owes the seller money. Once the actual settings are provided, there is no possibility for a gap to show up.

What goes for "owing" goes for "intending." If one considers it out of context, aiming for an "intending in itself," one is bound to create multiple intellectual problems instead of solving the one of elucidating intention. Indeed, disregarding circumstances might lead to a view of intention along the lines of the object-designator model and along the private-inner-object view that we've seen applied, and criticized by Wittgenstein, to sensations. On this view, intentions are inner states or processes. They are object-like in having causal powers. In particular, they can cause movements in animals and persons. Movements with intentions are actions; movements without intentions are reflexes or just movements (as when one is pushed). Intending is something essentially private. It goes on in the minds of persons and can always remain hidden, for the same movements can be made either without intention, or with a different intention, for example to deceive an onlooker into believing one has a certain intention one does not actually have. There is a principled gap between what people do and what their intentions are. Note how this view nicely fits with the privatization, objectification, and causalization of the mind that originated in the tradition of thinking about the mind exemplified in Locke and Hume (chapter 3).

Anscombe rejects this picture completely. Most fundamentally, she holds it makes no sense to think about intentions as "states" or "processes" without invoking the "forms of life" of animals and persons. These are the lives of creatures with needs and desires. Animals and persons want things and act to achieve what they want. Their lives and actions have a goal-oriented structure. They do some things to achieve other things. These doings often have a structure: One does C to do B to achieve A. And one can do further things D to do C, yet further things E to do D and so on. A cat might lower its back to stalk a prey unseen. A man might operate a pump to poison the inhabitants in a house. Anscombe uses such an example to illustrate the complex structure of that intentional action:

> A man is pumping water into a cistern which supplies the drinking water of a house. Someone has found a way of systematically contaminating the source with a deadly cumulative poison whose effects are unnoticeable until they can no longer be cured. The house is regularly inhabited by a small group of party chiefs, with their immediate families, who are in control of a great state; they are engaged in exterminating the Jews and perhaps plan a world war.—The man who contaminated the source has calculated that if these people are destroyed some good men will get into power who will govern well, or even institute the Kingdom of Heaven on earth and secure a good life for all the people; and he has revealed the calculation, together with the fact about the poison, to the man

who is pumping. The death of the inhabitants of the house will, of course, have all sorts of other effects, e.g., that a number of people unknown to these men will receive legacies about which they know nothing. (Anscombe, 1957, § 23)

Now suppose someone asks, with respect to either the pumping example, or that of the stalking cat, "Where is the intention?" A traditional answer is that the intention is a state, process, or, in any case, a "something" inside the person that is causally involved in producing the ensuing intentional behavior. The person has knowledge of that intention, but they could choose to keep it private so that it never be revealed. Within that same traditional line of thinking, one can choose either to deny that animals have such intentions, because they are cognitively not sufficiently equipped for that, or one might think roughly the same analysis applies to them. Anscombe disagrees. She claims that the intention is in the acting. But the acting is an acting-in-a-context, and many contextual aspects are necessary for there to be an intention. There must be forms of animal and human lives. A stalking animal must be hungry, needy of prey, and after potential prey. Humans must live in societies in which pumps are produced, operated by people to pump things like water; houses must exist with people living in them, people with evil plans must exist, meetings to conspire must exist, and so much more. On Anscombe's view, though many events and processes happen during the stalking or the pumping, such as overt movements or processes in the cat's or man's brain, none of them can be identified with "an intention." It is the whole situation, and the presence in the background of everything that constitutes a form of life, that allows for intention talk to be applied.

Anscombe claims that when we speak of events or processes as "mental causes," we do so in a different way than when we speak of intentions. That is, according to her, we do sometimes point to some mental event as the cause of an action (Anscombe, 1957, § 10–11). Suppose I ask you "Why did you suddenly jump up?" You might answer you did because you mistook a reflection in the window for an approaching object. Here you invoke some mental event that has just occurred to causally explain your action. With this explanation, the case is closed, in standard circumstances. But you could give the answer, "Because a process in my brain occurred that made my muscles move in a certain way" to my question "Why did you sign the petition?" Yet that would, in normal circumstances, not provide a satisfactory response. In offering an acceptable answer, in most cases, you do something different than pointing to some mental cause. You place the action in the circumstances and relate it to your role in those. To do that, you might say various things like: "I have sympathy for the originator of the petition, "I don't want to offend my friends," or "I want to prevent a new world war from happening." The three answers given would allow me to understand what you did, and that is something different from being referred to some inner event that mechanically caused the contractions of your muscles involved in putting your signature.

Anscombe admits that intentions can be, and sometimes are, kept private but denies that intentions are essentially private, like Wittgenstein did for sensations. That is, it doesn't make sense to conceive of intentions as always hidden, but it is possible, and sometimes actual, that people hide their intentions. If intentions were essentially private inner states or processes, they would, because of that essential privacy, be in principle unconnected to movements, actions and surrounding circumstances. Intending would be conceived of as "an interior movement" that would be totally free-floating (Anscombe, 1957, § 25). If intending is so radically disconnected from the circumstances that surround it, you could not have the "interior movement" and go on doing all the things you did. Or, you could have the interior movement and do nothing at all. In fact, this disconnect would not just be possible, it would be the rule. But on that picture, intending becomes ghostly, fully intractable, and completely unanalyzable. Nothing could determine what intentions someone had. Not even what a person said would count, because speaking is an intentional action too. To evaluate what a person says about their intentions, one needs to know what their intentions in speaking are. And if the current speaking needs more explication from the person, this brings in an additional round of speaking, that needs to be understood in the light of its intentions and so on to a next iteration, ad infinitum.

Now, one could ask: But doesn't the person know themself what they are doing? On a very traditional and intuitive picture, persons have a special kind of knowledge about their own intentions. They can know directly about them, while the intentions of other people are only known by observing what they do. But, on this standard picture, for knowledge of one's own intentions to come about, one does not have to observe one's own actions. Rather, one can look inward and have an inner perception of one's intentions. The view of intentions as inner states supports such thinking about a person's knowing about their own intentions. On this view, it is natural to think of the inner realm as something to which only the owner of this interiority has access to. Note that on this line of thinking knowledge of one's own intention is still conceived of in observational terms. The fact that the observation is of a special, inner kind does nothing to change that. Anscombe agrees with the traditional view that agents have a specific kind of knowledge of their intentions. That is, if one does not know what one is doing, one is not acting intentionally. If I am standing on a pipe and obstructing the water, but without any awareness of the fact that I am standing on the pipe, I cannot be standing on the pipe or obstructing the water intentionally. Yet, the knowledge is not observational in *any* sense. If one is acting, one's action is what is happening. However, one does not know what one is doing by observing what is happening, but by *doing* what is happening. How do I know I am typing now? By typing. If someone observed me and asked me "What are you doing?" I would of course answer: "I am typing," or "I am writing chapter 6 of the manuscript for *Of a Different Mind*." Now, if that person would ask me: "But how do you know?" it would be very strange for me to answer: "Well, I look at what my fingers are doing and from that I know that I am typing. And from reading what I am typing, I find out that I am writing chapter 6 of *Of a Different Mind*." Rather, I would be baffled by the question, and the only sensible answer I could give would be "But typing (or writing the sixth chapter) is what I am doing." Or so one should consider the case, according to Anscombe. Knowledge of intention is a special kind of nonobservational knowledge in which what happens is what one does, and what one does is what happens (Anscombe, 1957, § 29).

Intentional actions are structured: One does A to do B to do C, and so on. If agents have practical knowledge of their intentional actions, it seems they must have practical knowledge of this structure too. But what does it mean to say that they have knowledge of this structure? A way of answering that question is to think of this structure as something that the agent holds before their mind's eye. This approach is opposed to the overall thrust of Anscombe's thinking, because it seems to construe knowledge as an "interior movement" or inner act, something she argues against in the case of intention. The same reasons to reject intending as some inner act are applicable to knowing as an inner act. Indeed, she calls the interiorizing approach "absurd" (Anscombe, 1957, § 42). But Anscombe rejects the picture of practical knowledge as something standing before the mind's eye also for reasons that are specific to practical knowledge. If one construes practical knowledge on that picture, one assimilates it to speculative knowledge, and one loses, in one's theorizing, track of a difference that matters very much in life. If one models practical knowledge on a certain kind of contemplative knowledge, the special nature of practical knowledge is not recognized.

Anscombe's view seems to be that one practically knows the structure of one's intentional actions by being able to do them in a structured way. The cat has practical knowledge of stalking by lowering its back because it effectively stalks-by-lowering-its-back. Similarly, the poisoner-by-pumping has practical knowledge of poisoning-by-pumping because they *are* poisoning-by-pumping. The person is competently engaged in a structured action that they continue to produce because it has the structure they are producing: because that is what, in the agent's form of life, for agents that lead this form of life, it is to achieve something—poisoning—by doing other things—pumping.

Of course, the agent needs to be familiar with this structure for it to affect its actions. But to picture this familiarity as the contemplation of an inner picture of this structure is precisely to disconnect it from the circumstances that surround it, from the life of animals and the forms of life that need to be invoked in the Anscombean/Wittgensteinian approach to start making sense of concepts such as "knowledge."

This can be applied to the stalking cat. Rather than deriving from an inner picture, it is the circumstances that determine whether the cat's movements can be characterized as doings that exhibit practical knowledge. To be doings, the movements involved in the stalking of the cat must be attuned to the specifics of the circumstances and the overall context of the cat's activities, for example, to what it perceives and to what it will do if it catches the prey (Anscombe, 1957, § 47).

Anscombe is aware that there can be intentions, and knowledge of them, before one engages in the intentional actions that bring these intentions about. I can have the intention to poison the inhabitants of the house by pumping long before I start pumping. I could know that, and confess my intention, in private or in public, by saying, or thinking, "I am going to poison them, and I will do so by means of pumping poison into their cistern!" Isn't this a case in which something like the traditional view of observing inner states applies? This observational view is blind to something typical for expressions of intention: some sort of commitment to what one intends to do. If one says "I intend to go to the movies tonight," one does not take the stance of a neutral observer or reporter about what actually happens or has happened, but one commits oneself to a course of action. This commitment is something the person does, not something the person observes inside themself. One expresses one's engagement to make certain things happen by doing certain things. As already noted earlier in this chapter, expressing one's commitment is different from factual reporting. It is also different from factual predicting, even if it has a predictive aspect. This difference between factual and intentional prediction can be seen by contrasting the two senses of someone saying, "I am going to fail the exam." The most straightforward reading is to see it as a prediction of the facts. Someone realizes they have not achieved sufficient understanding, with no time left for remedy, to pass the exam successfully. But this sentence could also be an expression of intention. Someone could want to fail for the exam, perhaps because they would see that as the only way to be allowed by their parents to take a different subject of study. Now, if the person does pass the exam, the sort of error they made is different depending on whether the saying was a factual prediction or an expression of an intention. In the factual case, the person could have been mistaken about the facts. They could have been in error about which parts of the textbook had to be studied or about the level of difficulty of the exam. But in the case of expression of intention, the error would be one of having had a change of plans, or of having been insincere—perhaps the person tries to mask their lack of effort by portraying the failure as intended.

Anscombe would not deny that some uses of "I intend that . . ." might be used to describe one's intentions. For sometimes, for example, when people are puzzled about what they see us doing, they ask "What are you doing?" The agent might explicate their actions by invoking the broader intentional context: "I am crossing the road to read what's announced by the poster of the shop, because it might be about the festival I want to attend." These descriptions, or explications, elucidate one's actions by relating them to the wider intentional context. The actions are intentional by being inscribed in the intentional order and the explication renders their place in the intentional order visible.

For a great many of the actions we take, such elucidation is not necessary. One doesn't need to say, dropping an envelope in a postbox, that one is releasing this piece of paper to fall into this container, so that a person will come to collect it there so that it can be brought to a building where lots of pieces of paper that were dropped in a container are brought and sorted, so that they can be delivered to the place that is written on the surface of the covering paper in which I have put another paper. No, you just say you post the letter. The word "posting" presupposes this whole background. Our familiarity with the circumstances makes explication unnecessary and otiose. Much of the normal structure that forms a stable background for our intentional actions is presupposed in the ways we talk about actions. Words such as "posting," "replying," "stalking," and so on presuppose the presence of an intentional order.

To say that the word "posting" refers to something that could be described without invoking that whole background is to make a profound mistake, and the same applies, a fortiori, to "having the intention of posting." Now we can see why, on Anscombe's philosophy, the idea that "the intention of posting a letter" could be understood by studying an event, or process or state within the person, perhaps a neurophysiological one, is forlorn (Anscombe 1983). She rejects that the event, process, or state could be understood first, in a way independent from the circumstances, and that then this prior understanding would form the basis of understanding the intentions of the person. In her view, the reverse is true: One has to understand the activities in all their complexity, which include the wider temporal, social, and personal circumstances, before one can understand what the event, process, or state contributes to this overall process. In other words, the understanding of the event, process or state depends on the understanding of the circumstances, and not the other way around (Anscombe, 1983).

Humans and animals share some circumstances, but other circumstances are uniquely human. No animal can have the intention to marry or act intentionally to marry. Moreover, only a human who is familiar with the institution of marriage and knowingly participates in it can intentionally marry (Anscombe, 1957, § 48). Some intentional concepts like telephoning, greeting, signing, paying, or marrying can be applied to an agent's actions only if they are done with an awareness on behalf of the agent that they are engaging in that kind of activity—a kind of awareness that need not be present in other actions, such as intruding, kicking, or abandoning (Anscombe, 1957, § 48).

How does this all square with what motivated Anscombe to undertake her investigation of the concept of "intention," namely, her opposition to utilitarian ethics? Anscombe thought that a sound piece of philosophical psychology was needed to fight the justification that utilitarian philosophers gave for the nuclear bombing of Hiroshima and Nagasaki at the end of World War II. The issue was hot in her times because the University of Oxford in 1956 had decided to award Harry Truman an honorary doctorate (Mac Cumhaill & Wiseman, 2022, prologue sketches this background). Truman had been the president of the United States and thereby had the final word in the decision about the atomic bombing. For Anscombe, Truman could not be awarded a PhD because he had intentionally murdered hundreds of thousands of people. But utilitarian ethicists would disagree with Anscombe. They would argue that by bombing Japan, Truman prevented an even greater evil, with many more deaths than the bombings, namely, that Japan would continue to fight. In a utilitarian analysis, Truman comes out as having done a good deed, because of his overarching intention to end the war. In light of this wider intention, the bombing ceases to be a bad act of murder. Anscombe's analysis of intention refutes this conclusion. In her analysis, Truman did intentionally mass murder the inhabitants of Hiroshima and Nagasaki. This was because of what he did: dropping the bombs knowing that the inhabitants would be slaughtered. That he intentionally murdered those people is not changed by anything that he could have been thinking or saying to himself, such as "My real intention is not to murder these people, but to end the war." If a utilitarian analysis does not ascribe the intention to kill the inhabitants to Truman this just disqualifies the analysis. In Anscombe's own words: "While we can find cases where 'only the man himself can say whether he had the intention or not', they are further limited by this: he cannot profess not to have had the intention of doing the thing that was a means to an end of his" (Anscombe, 1957, § 25). So, even if his ultimate goal was to end the war, he pursued this goal by intentionally mass murdering the inhabitants of the two cities.

6.5 Ryle: Dualisms in Mind and Understanding

6.5.1 Ryle's Position

By focusing on sensation and intention in the foregoing sections, we have dealt only with a fraction of the mental concepts that Wittgenstein and Anscombe analyze and try to elucidate. However, the general contours of their approach to the mind have become visible. They both endorse "bringing back to life" as a philosophical method. Applying this method to language in the first place shows the limits of a picture-picture, or a referential view, of language. In applying this method and the ensuing view of language to the mental, they look for how the language of sensation, perception, intending, but also thinking, understanding, expecting, and so on, acquires shape in the complex forms of life of humans. Neglecting the context and circumstances in which these words occur leads to untenable impoverished views and to a lack of understanding. Instead of providing an increase of focus and the possibility to gain clarity, attempts to isolate the mental from the circumstances in which it occurs simply blinds us to the complicated and deeply contextualized ways in which minds exist.

Gilbert Ryle (1900–1976), professor of philosophy at Oxford, shared the philosophical outlook and attitudes of Wittgenstein and Anscombe. From early on in his career, he argued against a view of language as modeled on the object-name scheme—named "Fido-Fido principle" by him (Ryle, 1949b). Like Wittgenstein and Anscombe, Ryle insisted the mind is not, and cannot be, essentially hidden or private, even if people can lie about their experiences or hide their true aims.

In his most famous work, *The Concept of Mind*, Ryle offers a wide-ranging view of the mind (Ryle, 1949a). More than Wittgenstein and Anscombe, he tries to systematize and, besides offering detailed analyses of mental concepts, paint a broad picture of the mind. In one very conspicuous move he makes in the book, he very explicitly attacks dualistic thinking about the mind. Though addressing Descartes's substance dualism, Ryle's target is broader than that, for on his analysis, dualism extends much beyond the writings of Descartes in philosophizing about the mind.

6.5.2 Exorcising Ghosts

On Ryle's analysis, Descartes, upon being convinced that it is impossible to explain the mind mechanistically, invented a ghostly duplicate of the material universe that was at the same time essentially different from and yet in crucial respects similar to the physical realm. Descartes, in one of Ryle's famous phrases, endorsed the "para-mechanical hypothesis." According to it, next to a physical world, there is a mental world. Just as there are physical objects, there are mental entities, such as thoughts or perception. Just as events happen in the physical world, they happen in the mental world. Just as a physical object can causally impact on another physical objects, a thought can affect another thought. And, most generally, just as there is physical substance, there is mental substance.

Though many philosophers, from Descartes's times on, have rejected Descartes's substance dualism, Ryle thinks that strands of dualistic thinking have remained in much philosophical thinking about the mind. For a picture of the mind as essentially private and essentially inner has remained what Ryle calls the "official doctrine." This view is based on an opposition between "behavior" and "mind." It holds that the mind must be more than behavior, construed as the observable activities of persons or organisms. A prominent reason for the official doctrine is that it is always possible to behave in ways that seem to manifest mentality, while in fact nothing mental is actually present. That is, one can act as if one intends something, believes something, has had an experience one has not had, or has not had an experience one has had. This allegedly principled gap between behavior and experience shows, on the official doctrine, that the mind is essentially inner and private. It is something behind behavior that can be shown in behavior but need not be. Ryle refers to this view of the mind as "the Myth of the Ghost in the Machine."

This view of the mind as essentially inner and private is one that Ryle again and again criticizes. Like Wittgenstein and Anscombe did, he points out that minds should be understood in the broad temporal, social, situational circumstances in which they occur. In an elaborate discussion of "intelligence," for example, Ryle argues that whether or not someone or someone's action is intelligent is, normally, a matter of what a person does. But "what the person does" must not be construed narrowly. It is certainly not restricted to the movements a person makes now, but might involve long lasting stretches of the person's history and might very much depend on the current circumstances. Losing one's balance and tumbling over might be seen as a showing foolishness or drunken stupidity, but when a clown makes the audience cheer by tumbling exactly at the right moment, playing with expectations and exploiting patterns in their current performance and innovatively applying the format of clowning acts, intelligence is *in* the clown's performance. Yet when the clown repeats that same trick again and again, it becomes lackluster and dull.

A crucial point Ryle makes, again fully in tune with Wittgenstein and Anscombe, is that the mental cannot be *essentially* private and inner. If it were, it would no longer be connected meaningfully to activities; it would become a free-floating "interior movement" (in the words of Anscombe), a something that could even be totally absent, or "a wheel that turns nothing" (in the words of Wittgenstein, 1953, § 278). Against essential privacy, Ryle invokes the logical impossibility of essentially valueless money, or essentially forged money (Ryle, 1954). To understand what false money is, one must understand what real money is. Logically, understanding cannot operate in the reverse direction, because the concept of forged money is derived from the concept of genuine money.

Ryle famously often formulated his views in terms of dispositions about the mind. To say that person P has a disposition D is to say the person has tendencies to do certain things in certain circumstances. Applied to intelligence, Ryle proposes that whether the term "intelligent" applies to a person, or to a person's actions, depends on what the person does and is disposed to do. A performer can be called an intelligent clown if they do not just occasionally but repeatedly make the audience smile and say "ooh" by skillfully shaping, steering, confirming, and thwarting their expectations. The performer can be called an intelligent clown even when they are not engaged in clowning but currently dozing in their couch. They remain an intelligent clown even then because of what they are disposed to do in other circumstances.

But how should one understand these dispositions precisely? Carnap showed one way in his treatment of "excited" (see chapter 5). He modeled "excited" on the physical concept of "firmness." In his analysis, the firmness of an object should be conceived of in terms of the presence of a specific internal physical or molecular structure that explains, for example, why the object does not yield when it is pressed. Here one thinks of dispositions as *structural grounds* of activities. One explains the observed firmness of the object, understood as ways in which it reacts in certain circumstances, in terms of a structural feature—for example, in terms of closely connected components. Carnap applies this same analysis to "excited," proposing that it could be understood in terms of some structural property of the brain.

Such a view of dispositions is incompatible with the general thrust of Ryle's opposition to the Fido-Fido principle. The view of dispositions just discussed by reference to Carnap's example portrays dispositional terms as in fact referring to some underlying structure that explains the truth of certain statements of the form, "if *this,* then the object reacts in such and such ways." But to provide this analysis of dispositional terms is to conform to the

Fido-Fido principle, or, to use Wittgensteinian terms, to the designator-object model (Wittgenstein, 1953, § 293), which Ryle, as much as Wittgenstein, rejects. Further, this view of dispositions as pointing to grounds for reactions or behaviors, still seems subject to Ryle's criticism of the "two-worlds view." For it attempts to go beyond the activities, to point to something deeper behind the activities as the real object of terms such as intelligent. It differs from Cartesian dualism by conceiving of this "real object" in material terms, as a physical or neurophysiological structure. Yet, it still operates on an opposition between the actions of persons and their mentality. It is this opposition that Ryle's attack on the ghost in the machine is meant to target. The way to overcome the opposition, according to Ryle, fully in line with Wittgenstein and Anscombe, is not to point to something behind activities, but to more activities, and to their wider temporal, spatial, and social context. It is the possibly very broad context of the lives of persons, the forms of life of animals and humans, that make qualifications such as "intelligent" applicable.

So, on the Rylean view, we use words like "intelligent" based on a broad consideration of the activities of a person or animal. One way of interpreting the connection between the word "intelligent" and the circumstances on the basis of which it is applied, is to say that "intelligent" *means* "in circumstances C, person P will do A, in circumstances D, person P will do B, and so on." The meaning relation then is an analytic relation, as explicated in chapter 5. On such a view, "intelligent" is *defined* in terms of circumstances, just as one defines "circle" in terms of the condition that all points are equidistant from one given point.

But again, this view of the relation between terms and their circumstances of use rests on the Fido-Fido principle. It takes the fact that words are meaningful, to imply the existence of identifiable "meanings"—things that one could point to and say, "That is the meaning." But it is not necessary to think of circumstances of use as "the meanings" of terms. Rather, one can see them as conditions that need to be present for meaningful word use to be possible. In the ways in which we use words in certain circumstances, word use is meaningful. It is exactly this use-view of meaning that forms the basis of how Wittgenstein, Anscombe, and Ryle think about the mind.

Ryle famously coined the term "category error" for mistaking the function of a word. At the beginning of *The Concept of Mind,* he tells the story of someone being led around a university campus (Ryle, 1949a, chapter 1).

The person is shown various buildings: lecture halls, the dean's office, the administration tower, and so on. At the end of the tour the visitor might say: "It's been great to see all these buildings, but I still have not seen the university!" This person makes what Ryle terms a "category error," by wrongly assuming that the word "university" is just like the phrases "lecture hall" or "dean's office" in referring to a particular concrete entity. In reality, the word "university" functions differently. It obeys a different logic from the names for buildings. According to Ryle, the "ghost in the machine" view is based on precisely such a category error. It mistakes the actual variegated functions of mental terms for one of referring to some mysterious process behind the activities of a person or animal.

6.5.3 Knowing-How and Knowing-That

One prominent strand in Wittgenstein's work concerns rules (Wittgenstein, 1953, § 138–242). Much of what we do conforms, strictly or less so, to rules. We have rules for human interaction, rules for taking turns in conversations, rules for speaking grammatically, rules for telling a story, rules for writing poetry, and rules for solving a mathematical equation, to give only some examples. What we do can be right or wrong with respect to such rules. But how can such sensitivity to rules be explained? Wittgenstein criticizes certain ways of conceiving of such explanation. In particular, he discards the possibility that rules are mental entities that cause rule-guided activity. On such a view, when you speak a sentence that is well-formed, you apply a rule that is stored somewhere in your brain. The rule prescribes how you should form sentences, you consult it, and then you form your sentence. Of course, if one holds such a view, one might add that the rule and its consultation can remain unconscious.

Wittgenstein rejects this way of explaining rule-guided activity, because, among other things, it doesn't succeed in explaining rule-governedness. All it does is replace rules in activities by rules in the mind. But what makes it possible for you to follow a mental rule is not explained. Just the appearance of a rule before the mind's eye doesn't suffice. If I see a real rule in a language I don't understand, I will not be able to apply it. The same goes for a mental rule appearing before my mind's eye. Applying a rule requires understanding or interpreting the rule. Such understanding is not random but follows rules itself. Now, remember that the first rule in our example was invoked to explain one's understanding of grammar. But if the first rule requires understanding, then, if one applies the initial reasoning, this requires a further rule. But then this second rule needs to be interpreted again, which requires a third rule, which, on the same logic, needs to be interpreted again, and on and on. An infinite regress occurs. Perhaps one can evade these worries by saying that the mental rule needs no interpretation, but just automatically gives rise to the right behavior. But now one has deprived the mental rule of any explanatory power. If there is no interpretation, the application of the rule just becomes acting according to the rule. The rule has become identical to the occurrence of the rule-sensitive activity that it was supposed to explain. One says that one constructs the sentence grammatically because one constructs the sentence grammatically. That is to offer no explanation at all.

But if a mental rule doesn't, what then explains the correct use according to Wittgenstein? It should come as no surprise that Wittgenstein points to the circumstances that surround the regular use: the practices of use in a form of life. Rules are aspects of such practices; it is where they emerged and where children acquire familiarity and competence with them. This happens in a gradual process of learning that exploits existing but different capacities in the learner: capacities for being sensitive to similarities, or to react to corrections or praise, for example. Learning the rule is acquiring the capacity to participate competently in the practices. To think of this learning process as the formation or internalization of some mental rule would be to run into the initial problems with mental rules. One could have a rule before the mind's eye and not know how to put it into practice, just like one could have a musical score before the mind's eye and not be able to sing or play the music. Of course, the Wittgensteinian view has the duty to explain how rules can emerge in practices—otherwise, it has simply replaced the mental rules that it rejected because they were unexplained explainers by other unexplained explainers: rules in the practice. The thrust of Wittgenstein's thinking here is to apply the overall method of "bringing back to life." To understand why there are rules in practices one must return to the circumstances, including the "natural history" of how human forms of life in which regular practices figure originated from simpler forms of life in which they didn't. In this way, rules in a practice are not end points in the way that mental rules are: They can in principle be further explained, while mental rules, as criticized by Wittgenstein, are meant to be basic.

Ryle famously has attacked what he calls "the intellectualist legend" about the mind. The overall point Ryle wants to bring home is that intelligent action is not and cannot be based in theoretical thinking. Rather, it is the other way around: Theoretical thinking is a specific form of intelligent, nontheoretical thinking. Ryle's discussion of the intellectualist legend rests on a distinction between knowing-how and knowing-that. The distinction is motivated by considerations very similar to Wittgenstein's views on rules. Knowing-that is theoretical, explicit knowledge. One has knowledge-that of a period in history if one can recite facts and tell narratives or write a book about that period. Knowing-how is practical: One knows how to tie one's shoes, because one can do it. In many cases, one cannot make one's know-how explicit. Most people can tie their shoes, but they will not give a correct answer if they are asked to describe, on the spot, with no shoes to tie near them, what exactly they are doing when they tie shoes. But this inability to make explicit one's knowledge doesn't at all form a reason to deny that someone has the requisite know-how.

The intellectualist myth Ryle wants to disband is that knowing-that forms the basis of knowing-how. If that were the case, one would have to theorize about how to do something before one actually did it. Before one tied one's shoes, one would have to devise a plan of how one would grasp the ties, in what order one would move them around, how and in what direction one would pull, and so on. Of course, we don't engage in such thinking consciously, but perhaps, it could be said, we do so unconsciously? Ryle points out a problem that arises for both conscious and unconscious knowing-that. It invokes an infinite regress—very similar to the one discussed in the context of Wittgenstein's treatment of mental rules. The knowing-that-first approach assumes that conscious or unconscious explicit planning lies at the basis of intelligent doing. But, so Ryle points out, such planning can be done in two ways: intelligently or not intelligently. If it is not done intelligently, it cannot form the basis of intelligent doing, because it would lead one to do the wrong things. So, it must be done intelligently. But if that is the case, then, if knowledge-that-first is correct, the planning required to make the action of tying intelligent, itself requires another explicit, knowledge-that basis. One has to plan how to do the planning of the tying for the planning of the tying to be intelligent. Now, what was said about the planning of the tying applies to the newly invoked round of planning it too. Another round of planning of planning appears, to which the whole cycle of reasoning applies again. One cannot stop to invoke further rounds each of which demands one more chain preceding it. One is off to an infinite regress.

Such infinite regresses are avoided if one inverts the intellectualist picture and takes knowing-how to be the basis of knowing-that. Knowing-that then becomes a special form of knowing-how: knowing how to theorize, to explicitly plan, to state and relate facts, and such things. Ryle illustrates this inversion by considering logic (Ryle 1949a, chapter 2). Nowadays, logic has become a self-standing academic discipline. Important parts of logic concern propositional logic and predicate logic. Propositional logic deals with such facts as that one can conclude from "It is summer and it rains" that it rains, or from the truth of both "If he is in Paris, he buys perfume" and "He is in Paris" that it is true that he buys perfume. Predicate logic is about the validity of reasons such as drawing the inference that "Socrates is mortal" from "All humans are mortal" and "Socrates is a human." Propositional and predicate logic both partially consist of schemes of inference corresponding to such examples. Ryle remarks that before such systematic logics were developed, people already made the inferences that were later codified in the logic. Before Aristotle developed his predicate logic, people already made inferences like in the Socrates example. A fortiori, such is true for propositional logic, which was developed many centuries after Aristotle. Logic is comparable to a detailed description one might make of tying a shoe—something quite difficult to achieve in a purely theoretical, verbal way, without making use of diagrams. Both come necessarily after what they theorize about already exists. Therefore, they cannot form the basis of practices of tying or reasoning.

In his later work, Ryle thought a lot about thinking. Even for thinking he wanted to get away from an overly "theoretical" portrayal of thinking. On such a view, thinking is modeled on "having an explicit thought." To fight that picture, Ryle points to the many ways in which people can think. A sculptor might think about how to carve a large piece later by first forming and arranging small balls of clay. They might try some composition, inspect it, and then rearrange it. The sculptor thoughtfully plans the piece, but we need not assume that something like a citable thought occurred in their mind. And even if someone would be talking to themself, and could say what they thought, the "sentence in the head" would amount to thinking only in certain circumstances, such as when they would attempt to change a course of action or was dealing with something they find puzzling (Ryle, 1968/2009). There is much in common between how Ryle thinks about "thinking" and how Anscombe thinks about "intention." Instead of asking "What is a thought?" or "What is an intention?," the question becomes: Under which circumstances do the words "thinking" or "intending" apply? The answer to that question consists in what animals and people do in certain conditions, and, importantly, the way they do things. To reframe the questions about thought and intention in these ways leads away from the tendency to look for the answers in terms of the occurrence of some mental object, or, for that matter, some mental process, event or state. Importantly, it equally breaks with thinking about thinking as a process. Rather, it invites to think about thinking in terms of ways people do things. Thinking is not a state or a process, but a certain manner in which actions are engaged in, and one can say similar things about intention. Unsurprisingly, one finds the same attitude explicitly in Wittgenstein. For example, he writes the following: "Now when one says of a human being that he has a mental life: he thinks, wishes, fears, believes, doubts, has images, is sad, merry etc.,—is that analogous to: he eats, drinks, speaks, writes, runs,—or analogous to: he moves now fast, now slow, now towards a goal, now without any goal, now continuously, now in jerks? (Wittgenstein, 1980a, § 284).

Steiner (2018) suggests that Wittgenstein opts for the second answer. The common attitude of Wittgenstein, Anscombe, and Ryle can then be characterized as some form of "adverbialism" (see, e.g., Steiner, 2018; also Ryle, 1968/2009, p. 481, though Ryle refrains from firm commitment to that term), as it relates mental terms to manners in which organisms and persons act in certain circumstances. Such adverbial approach to the mind stands opposed to what can be called a nominal approach—one focused on what names do, namely refer to objects, states, or processes. An adverbialist analysis of intention will characterize intending as a way of doing things, rather than as something relying or being caused by some inner object—an intention. It will not necessarily object to speaking of "an intention," but it will see such talk as not naming an inner object, but rather as characterizing which complicated circumstances allow the application of phrases containing "intention."

Bibliographical Note for Chapter 6

The literature on Wittgenstein is massive, even if one would restrict it to his work on the mind and psychology. A book I found particularly insightful is Michel Ter Hark's *Beyond the Inner and the Outer: Wittgenstein's Philosophy of Psychology*. Several publications by Danièle Moyal-Sharrock might be found to be of relevance to the issues and strands of thinking treated in the current book (Moyal-Sharrock, 2004, on Cartesian doubt; Moyal-Sharrock, 2000, on Wittgenstein and contemporary philosophy of mind; and Moyal-Sharrock, 2013, on Wittgenstein and enactivism). Excellent books on Ascombe's *Intention* have been written by Rachael Wiseman and John Schwenkler (Schwenkler, 2019; Wiseman, 2016). Regarding Ryle, the work of Julia Tanney critically relates Ryle to contemporary philosophy of mind (Tanney, 2009 and Tanney, 2022).

Ryle's work on knowing-how as being more fundamental than knowing-that has been the object of intense critique since the appearance of Jason Stanley and Timothy Williamson's paper arguing exactly the opposite: that knowing-that is fundamental (Stanley & Williamson, 2001). For rebuttals of Stanley and Williamson, and the knowing-that first approach, see Noë (2005) and Hutto and Robertson (2023).

7 Cognition First: The Cognitive Revolution and Its Science

> The activities of the computer itself seemed in some ways akin to cognitive processes. Computers accept information, manipulate symbols, store items in "memory" and retrieve them again, classify inputs, recognize patterns and so on. . . . Indeed the assumptions that underly most contemporary work on information processing are surprisingly like those of nineteenth century introspective psychology, though without the introspection itself.
>
> —Ulric Neisser (1976, p. 119)

7.1 A Revaluation of Values

In chapters 2,3, 4, and 6 we have seen how Thomas Reid, John Dewey, B. F. Skinner, and Wittgenstein, Anscombe, and Ryle, each in their own ways, argued against what they considered to be misleading pictures of the mind. In their works, one finds arguments against the idea that perception relies on representation, that intentions are causal inner states, that experience is essentially private, or that cognition is to be explained by some kind of inner rule-following. However, roughly since the second half of the twentieth century, philosophers increasingly reasserted what the earlier-mentioned philosopers rejected.

To provide a first illustration, consider what one of the most cited philosophers of the last decades, Jerry A. Fodor (1935–2017) and his frequent coauthor Zenon Pylyshyn, write in their 2015 book, *Minds Without Meanings*: "In the paradigm cases, behavior is the effect of mental causes, and the paradigm of explanation in cognitive psychology is the attribution of a creature's actions to its beliefs, intentions, desires, and its other "propositional attitudes" (Fodor & Pylyshyn, 2015, p. 2).

What Fodor and Pylyshyn endorse is exactly what Wittgenstein, Anscombe, and Ryle fought against. For example, in the quoted passage, and more extensively in the broader philosophy Fodor and Pyslyshyn defend, they embrace a view of intention that Anscombe criticized for resorting to "interior movements" as fundamental and for failing to distinguish between intentional talk and talk about mental causes—to mention only two of the shortcomings Anscombe diagnoses in that view. Fodor, who took a prominent role in philosophy of mind and the emerging discipline of cognitive science from the 1970s on, highlighted his differences with the thoughts of thinkers in the tradition of Wittgenstein, Anscombe, and Ryle in many of his publications. In an earlier book, Fodor (2008, p. 10) expresses his opposition to philosophies in which it is claimed that "abilities are prior to theories," that "competence is prior to content," and that "knowing how is . . . prior to knowing that." Fodor's stance is the complete inverse: There are forms of theoretical thinking that precede the learning of language in social circumstances (Fodor & Pylyshyn, 2015, chapter 1). Or, a capacity for knowing-that is a preexisting cause for human practices in forms of life, instead of a product of such practices. As in the passage quoted earlier, Fodor is convinced that intentions, beliefs, and the like are internal states that cause behavior. Though Jerry Fodor has gone further than most in committing to the consequences of these theses, a lot of thinking about the mind since roughly the 1960s has gone in this same direction. Moreover, as will become clear in this and the next chapter, many other elements of the views of the mind held by thinkers such as Descartes, Locke, Hume, Müller, and Helmholtz became revived. Many mainstream thinkers, both in science and in philosophy, since the second half of the previous century endorse that experience is internal and essentially private, that perceiving the world consists of forming an internal representation of the world, and that such perception relies on processes of unconscious inference.

What made this revival possible? There were several converging causes, among which stand out a lingering dissatisfaction with what was seen as "behaviorism" and a fascination for everything having to do with "information" and with computers. In the 1950s, momentum gathered for declaring the arrival of a new scientific discipline called "cognitive science" (Gardner, 1985). Cognitive science was meant to be a science of all cognitive phenomena and to include several approaches: neuroscience, psychology, anthropology, linguistics, and philosophy. All those disciplines were taken to be unified by conceiving the mind as a "system" that was both representational and computational—in ways which will be further explained shortly. In fact, people working in the different disciplines constituting cognitive science did rely on each other. Cognitive neuroscientists or cognitive psychologists saw their work vindicated and justified by the work of like-minded philosophers and vice versa. The collaboration between Jerry Fodor, a philosopher, and Zenon Pylyshyn, a psychologist, forms an example of such cooperation within the different disciplines united in the grander cognitive science project.

7.2 Preparing the Ground

The cognitive science view of the mind as representational and computational did not appear out of thin air. Besides the strands of mediationist and representationalist thinking that have roots at least as far as Greek antiquity, there were more recent developments that favored the arrival of cognitive science on the intellectual scene.

One important relatively early development concerned logic and its formalization. Logic is a very old discipline. At the end of the eighteenth century, Kant could still claim that progress in the discipline of logic was not possible, because Aristotle had given the definitive treatment. Kant was proved wrong about his assessment when in the nineteenth and twentieth centuries, logic was thoroughly revised and expanded, in the work of people like Gottlob Frege (1848–1925), Bertrand Russell (1872–1970), and many others.

In the hands of Frege, Russell, and many others, logic was formalized. Means were found to transform reasonings such as:

If it is winter, then it rains
It is winter

———

Therefore, it rains

into sequences of symbols such as

p -> q
p

——

q

Further, rules were devised for moving from one symbolic formula to another that quite accurately matched what we consider a valid way of reaching conclusions from premises in everyday situations. Formal logic has a rule that allows to change the formulas

p -> q

and

p

into the formula

q

So, a logic can be devised in which the transitions allowed between symbolic formulas correspond to transitions in everyday reasoning. The formal mirrors the informal.

It is crucial that the rules of formal logic apply because of the form of formulas, not their specific meaning. "p" and "q" could mean anything—for example "It is winter" (for p) and "It rains" (for q). Anytime one has as a given something of the form "p -> q" and something of the form "p," one can arrive at "q."

The formalization of logic opened the prospects for the mechanization of logic. If an operation is formal, it depends only on combining variously shaped symbols and sequences of them into other symbols or symbol sequences. It does not require that one understand the meaning attached to those symbols or what the symbol sequences represent. A machine can produce the output "q" when given the input "p" and "p -> q" without knowing anything about the rain, the winter, or whatever. Because the rules of logic can mirror the rules of everyday reasoning (or at least to an extent), the fact that logic can be mechanized means that everyday reasoning can—at least to some extent—be mechanized.

How would such a formal reasoning machine work? A principled answer to that question was given by Alan Turing (1912–1954). Turing's work concerned the question of what exactly can be formalized in the way that formal logics formalize logic. To answer the question of how much of mathematics and logic can be formalized, the so-called "theory of computation" was devised. It connected the possibilities of formal systems to what can be achieved by formal operations—such as transforming one formula into another—in a finite number of steps. Such transformations were called "computations." Turing introduced the notion of a theoretical machine—an idea of a machine—that later was called "Turing machine," after its inventor. As said, computability theory concerns what can be calculated or computed in a number of steps. A very simple example of a computable problem is the "What is the number x plus 2?" problem. You can solve that problem for any given number by just adding two and writing the result. You get 1 and you produce 3, 5 and you produce 7, for example. More complicated computational problems require more complicated sequences of steps, or algorithms, of course. Moreover, some mathematical problems can be formulated that cannot be computed. Turing invented Turing machines to make the notion of computability more precise, that is, to lay down a definition for the relatively informal notion of "computable in a number of steps." Turing proved that existing formal notions of computation were equivalent to what can be computed by a Turing machine. He also proved that it is in principle possible to construct a universal Turing machine that can do everything any other Turing machine can do.

How does a Turing machine work? Interestingly, Alan Turing modeled his machine on a "man in the process of computing a real number" (Turing, 1937, p. 231). Turing machines consist of a tape, divided into squares, and a writing and reading device. On the squares can be symbols, one per square. The reading and writing device can recognize symbols written on a square, and it can write one symbol at a time. If it writes a symbol on a square that already contains a symbol, the old symbol gets overwritten: It disappears and makes place for the new one. Further, the machine can be in different "states." Think of these in a general way as conditions the Turing machine is in. All that matters for states is that they are different from each other. Further, the machine can move along the tape, to the left or to the right. What the machine does at any moment is determined by: (i) the symbol it is currently scanning with its reading head and (ii) the state it is in.

Suppose we want to build a Turing machine that adds two numbers. We write out numbers in a simple fashion, in so-called unary notation, as vertical stripes. So, we want the machine to render, or write, on its tape "IIIII" when it is presented on its tape with "III" and "II." How would we construct such a machine? We let it start on the left end of a tape on which are written two numbers separated by a blank and followed by an end symbol, "#." We want it to add a three and a two. Then we have three "I's," in three subsequent squares, followed by a blank, followed by two "I's" followed by the stop symbol. When setting up the machine we should consider the strategy it will follow to successfully accomplish the task. An efficient strategy is to let it move to the right over the three "I's," until it reaches the blank. It should replace the blank by an "I." Then it should go to the end of the second group of "I's." When it has reached the last "I," it should overwrite it with a blank. Then the machine can stop. It basically has removed the blank between the two groups of "I's," and then it has erased the final "I."

Clearly, the machine should act in the appropriate way after it has encountered the stop symbol, which is "#." It should "know" that it has reached the end and, therefore, should replace the "I" to the left of the "#" with a "0." The machine achieves this by switching from state q1 to state q2.

The workings of the machine can be described by a machine table. For our machine it looks like this:

	I	0	#
q1	Rq1	IRq1	Lq2
q2	0q2		

The "#" is the symbol marking the end of the numbers. The "0" marks a blank. If we assume the head of the machine is over the leftmost "I" at the beginning, the machine table tells us that the machine will then go to the right and, if it encounters an "I," will remain in state q1 (Rq1). If it encounters another "I," it will do exactly the same. If it then encounters a "0," it will replace it with a "I," move to the right and stay in q1 (IRq1). If it encounters a "#," it will move left and switch to state q2 (Lq2). Then it will be in q2 and read an "I." The machine table shows it will then replace the "I" with a "0," stay in its place, and stop, because the table contains no instructions for when it is in q2 and scanning a "0" (the one it has just written). One can think of the machine table as a blueprint of how the machine is made to run. Different machine tables then show different ways in which a machine can be set up or programmed.

Our example of a Turing machine is quite simple and performs a mundane task. But in principle, other tasks, such as deriving a conclusion from a set of premises in propositional logic, and many much more complicated feats can be performed by a Turing machine. Moreover, Turing showed that one can construct a universal Turing machine: one Turing machine that can do everything that all possible Turing machines can do. It basically works by treating the machine tables of other Turing machines as part of its input, and then simulates what they would do.

Turing machines are mathematical constructs; they are not actual machines. But it is possible to build actual computing devices. Nowadays we are surrounded by such things: desktops, laptops, mobile phones, and what have you. Though there are some architectural differences, our computers are similar to Turing machines. The tape of a Turing machine is whatever in a computer can be used to provide input and whatever can be used to write output. A state of the Turing machine corresponds to any different condition that the machine can be in that is designed to be relevant for its functioning. Your computer is in certain states when it is running different programs, and when it is doing different things within one program. One computer can also simulate another computer, just like a universal Turing machine can simulate another Turing machine. Unlike Turing machines as conceived by Turing, our computers don't have infinite tapes. Hard disks can store only so much data. On the other hand, there are ways to enhance the memory capacities of computers, by adding on devices, and there is no obvious in-principle limit to such memory upgrades. Crucially, Turing machines and our computers are machines that transform symbols into other symbols, in a regular, rule-based way. Digital computers code for all symbols in digital numbers so that a "p" could become some sequence of zeros and ones. Whether symbols are numerical or not doesn't matter, at bottom.

These developments in logic and in the theory and technology of computation went hand in hand with an increased intellectual interest in the notion of information. The word "information" has existed for a long time. In informal conversation, we can say that one newspaper provides more or less accurate information than another newspaper. Information, then, is "what is said." I can provide the same information in French as you did in English. In a related use, one can say that a weathervane provides information about the direction from which the wind is blowing. But in the twentieth century, increasingly the conviction grew that "information" could be turned into a scientific term, and a very important one at that. Perhaps it is useful to compare with the notion of force. Before the notion of force obtained its standard technical physical meaning in the work of thinkers like Galileo and Newton, it was a word widely used in the common language. Due to a horse's pulling force, the wagon moves, and when the horse is tired and no longer applies sufficient force, the wagon stops. Besides, both an army and an insult can be forceful. But the technical notion of force, apart from being much more restricted, was different from the everyday notion. Indeed, it is considered to be one of the great insights of physics since Galileo that physical forces are not necessary to keep motions going (Dijksterhuis, 1950). Forces are only needed to change—that is, accelerate, decelerate, or redirect—the motion of an object. Yet despite these differences, the physical notion of force can be considered to be the theoretical counterpart, refined to be sure, of the existing notion of force.

It seems a similar expectation lived in the twentieth century, and probably ever since, with respect to the notion of information. The celebrated work of Claude Shannon (1916–2001) is often described as having laid the foundation for a discipline called "information theory" (Collins, 2002). Shannon (1948), approached information as a signal that can be sent over a physical channel from a sender to a receiver. Using probability theory, he devised a quantitative measure for the amounts of information that could be conveyed in this way. Shannon's work played a major role in promoting the concept of information to a primary role on the scientific scene. "Information" became seen as a potentially or actually fundamental scientific notion, equal to "force," "matter," or "energy." Expressing this, Norbert Wiener, another scientist at the forefront of theories of information and computation, wrote: "Information is information, not matter or energy. No materialism which does not admit this can survive at the present day" (Wiener, 1961, 132. Cited in Gardner, 1985, p. 21).

"Information" became a common concept in biology (in particular, genetics), in physics (thermodynamics), and, of course, in computer science. It remains an open question, however, how these technical uses of "information" relate to the everyday notion of information as "what is told." It is not clear at all whether this common "semantic" understanding of information as "what is told" is retained in the technical uses (see Griffiths, 2001; Zahnoun, 2025). Overall, it is an open question whether the technical use of "information" stands to the common understanding of information as the physical concept of force stands to the informal notion of force that was around first. Despite this, the notion of information has become widely embraced in thinking about the mind.

7.3 The Cognitive Revolution and Classical Cognitive Science

7.3.1 The Overall Picture

These developments in ideas about computation and information, together with the complaint that behaviorism left important aspects of the mind out, gave rise to the emergence of cognitive science (Gardner, 1985). Cognitive science is about cognition. But what is cognition? In a restricted use, "cognition" would include phenomena like reasoning or planning, but not necessarily perception or movement. But in a wider use, "cognition" also covers perception and at least some movements. The drawing of the boundaries of cognition can be based on where one sees information processing or computation and representation at work. If one considers perception or motor activity as involving such information processing—a very widely held assumption—it tends to be termed cognitive, too.

A core idea of cognitive science is that cognitive phenomena come about because of the operations of cognitive processes. Cognitive processes are standardly conceived as processes inside a "cognitive system," which could be a person, an organism, or a robot. Typically, a cognitive process operates on information present in the system. In a cognitive process information could be compared to other information, or information could be written in and retrieved from storage places or memories. It could be copied and moved from one kind of memory to another. If information is numerical, it could be subject to arithimetic operations, such as addition or multiplication. If information is symbolic, it could be transformed from one symbol to the other. Cognitive operations include relating a piece of information to a stored template to see if a match can be found. They could consist of comparing newly incoming information to knowledge already present, so that either the new piece of information could be completed or interpreted, or the existing knowledge amended or updated.

From this gloss, it becomes visible that cognitive operations are similar to things persons do. Cognitive operations are thus conceived in the same way as Turing conceived of the operations of a Turing machine: on the model of a person engaging in activities like working with a pen on a page, dividing it into squares, writing on it, reading, and comparing what they now read with what they remember (Turing, 1937, p. 231). Showing that a machine could perform such humanlike operations is seen as one of Turing's major contributions (see, e.g., Dennett, 2009). Turing was very precise in describing how his imaginary machine would perform the analogs of the operations carried out by his "man in the process of computing a real number." The subsequent treatment of "cognitive operations" aims to achieve the same level of precision by conceiving of cognitive operations as "computational." They are meant to be operations that a computer can do.

Cognitive operations are standardly taken to operate on representations—that is the way they process information. In thinking, remembering, perceiving, and other cognitive capacities, cognitive systems represent the world as it is, or as it could be. A representation is something that specifies, pictures, or describes something else. Thereby representations have "aboutness" or "intentionality," a notion that will be discussed in more depth in chapter 9. Aboutness is the property of "reaching out" to something else, something physically outside of the representation, yet in some other sense "contained" in it.

"Representation" is a core notion of the standard cognitivist approach to the mind (Gardner, 1985). Cognition is all about representing. In perception, cognitive systems receive information from the outside world so that perceptual representations can be constructed. Such representation can be stored inside the cognitive system, to be retrieved later as memories. Thinking operates over representations that can have their origin in perception, in memory, or in other cognitive capacities, including thinking itself, to lead to new representations. For example, in planning (a form of thinking), a cognitive system combines representations from perception, memory, and volition to come to a new representation: a plan. Once generated, the plan will guide action: It will serve as a blueprint representing the goals to be achieved and possibly also the ways to achieve those goals.

So, cognition is information processing understood as representations that get operated upon by cognitive or computational operations. Typically, all of the computational and representational activity is taken to occur, for humans, within the brain. The idea that cognition fundamentally happens within the brain is called "internalism about the mental." Classical internalism about the mental combines two ideas. First, for cognition, what matters is how the world is represented. Second, representation happens in the brain (see Fodor, 1980, for a statement and defense of such internalism).

The cognitivist approach thus rests on three pillars: computation, representation, and internalism. By and large the cognitivist approach reconnects with what Reid (1764) called "the way of ideas"—the view of the mind held by the British empiricists à la Locke—and with scientific psychology as practiced by, for example, Helmholtz. Like in these traditions, the cognitivist approach is mediationist and representationalist. The cognitive operations of the cognitivist can be seen as related to the processes of comparison and association one finds in Locke or Hume, or the processes of inference one finds in John Stuart Mill, Helmholtz, and many others (see chapter 3). A crucial innovation is that the mental operations of the cognitivist are taken to be computational. This is important because construing cognitive operations as computational is taken to give them scientific credibility. If a cognitive operation is a computational operation, it can be accepted in a scientific or naturalistic worldview, in which nothing exists besides natural entities and processes that can be understood scientifically. Seeing cognitive operations as computational allows cognitivists to reject Descartes's explanatory reasons for dualism. As seen in chapter 2, Descartes's dualism was in part motivated by the thought that reasoning could not be mechanized. If thinking is computation, however, then, since computation, as Turing has shown, can be mechanized, thinking can be mechanized too (Clark, 2014, chapter 3; Dennett, 1978; Fodor, 1983).

So, though philosophers and psychologists had speculated intensively both about mental entities (ideas, sensations, percepts) and mental laws or rules that apply to them, they did not really show how the laws of the mind—for example, Hume's principles of association or Helmholtz's operations of unconscious inference—should be thought of mechanistically. With Turing machines and subsequently computers, it became possible to imagine the dynamics of thought in a detailed mechanistic way, namely, as resulting from simple operations on input inside a cognitive system.

Thinking about mental operations as computations was also taken by cognitivist philosophers to evade the problems raised by Wittgenstein and Ryle concerning rule-following. Cognitivist theorists held that stored knowledge, plans or programs can guide action and direct cognition in rule-governed ways. Wittgenstein and Ryle had argued that assuming mental rules, recipes, or playbooks to play the role of guiding knowledge and action runs into unsolvable problems. For example, they argued that rules require they be understood before they can be applied, and understanding is, again, a rule-governed activity. But if rules require new rules, these new rules require yet more rules, and so on without end. Cognitivist theorists claimed their invoking of plans, programs, and other rule-like entities didn't run into this problem because they think of following a plan or being guided by a rule in computational ways (Dennett, 1978). To follow a rule, so they will say, is just to perform certain computations and not others. Wittgenstein did argue that such mechanistic conception did not explain the human capacity to follow a rule, but rather restated the capacity in a mechanistic language, but this further consideration didn't prevent the cognitive revolution to happen (for the relevant views of Wittgenstein on rules, see this book, chapter 6; also Thornton, 1998).

An important reason to see a Turing machine, or more generally a computer, as a model for cognition was the generative or systematic nature of its operations (Dennett, 1987; Fodor & Pylyshyn, 1988). To understand this, consider again the versatility of human rationality or cognition that formed an important reason for dualism, from Greek antiquity to Descartes. For example, both Aristotle and Descartes contrasted what they saw as a human capacity for learning anything, seemingly without limit, with the ability of animals to acquire specific patterns of responding (see chapter 2). Without necessarily claiming that animals lacked this generative capacity, cognitivist philosophers ascribed it to the brain. For example, Daniel Dennett (1987, p. 35) characterized the cognitivist picture of the brain as follows: "It is a gigantic network of billions of cells, but still finite, compact, reliable, and swift, and capable of learning new behaviors, vocabularies, theories, almost without limit."

Cognitivists have argued that the brain could be the basis for such generative flexibility if it operated like a computer (Fodor & Pylyshyn, 1988, and, more cautiously, Dennett, 1987). Even a simple Turing machine can deal with an infinity of inputs, and will, given that they have the right shape, always provide the right output. A well-programmed computational system will give the right answer to "What is x + y?," whatever numbers are filled in for x and y. Cognitivists like Chomsky and Fodor have claimed that human cognitive capacities show this generativity because an underlying computational system is involved in the causation of those activities. The human cognitive capacity for producing and understanding an infinite amount of sentences in natural languages, according to Chomsky or Fodor, is possible only because some computational system is involved in the processes of production and understanding. Ultimately, human cognition rests on a system that already has some of the fundamental characteristics of cognition. Cognition comes first.

In summary, embracing the view of cognition and mind as computational and representational looked like a way to at the same time abandon discredited ways of viewing the mind, renew contact with longstanding approaches to the mind, and reinvigorate them with new scientific ideas about information processing, computation, and representation.

7.3.2 From Vision to Language

One of the most celebrated classical pieces of work in cognitive science is due to David Marr (1945–1980). Marr's work was immensely influential, both because of its methods and its specific content.

Marr's account of vision (Marr, 1980) starts from the idea that vision has a task, namely, to make information available about the layout and spatial position of the objects in an environment. Very much in the spirit of the overall cognitive science project, Marr conceived of the accomplishment of this task in terms of building representations that made explicit this layout and spatial position. Very much in a Helmholtzian spirit, Marr stressed the need for processing the input received by the visual receptors—the visual "image" projected to the retina at the back of the eye after light entered the eye and passed through its lens. As Helmholtz had made central to his approach to perception, the relevant information about form, depth, and other aspects seen was present in the stimulus only in degraded and ambiguous ways. For example, the retinal image is fragmentary: It shows only how much light is sent to the retina from various small parts of the world. Moreover, in the human visual system, two eyes together looking at a visual system from a different angle lead to two retinal images. For seers to become aware of which fragments of the retinal image belong together by being of the same object, the retinal image needs to be operated upon, so that a representation is created that explicitly represents the contours of objects. Computations are also required to construct a representation that makes explicit the three-dimensional layout of objects and scenes by operating on and comparing the left and right two-dimensional retinal image. By means of finding corresponding areas on both retinas—those that are stimulated by the same place in the outside scene—and by comparing their different position on the retina, the three-dimensional layout of the scene can be reconstructed by computational processes in the brain. The process, called "stereopsis," is reminiscent of how Descartes explained seeing depth (as described in chapter 2).

Marr proposed that the brain contains several specialized subsystems, each dedicated to aspects of vision, such as stereopsis, motion, or color. Such systems came to be called "modules." Each module applied specific computational operations specially tailored to its area of expertise. The form system computed contour from patterns of light changes to create something that can be visualized as a contour drawing. The motion module computed direction and velocity of motion by computing representations of motion from temporal patterns in the visual input. Several layers of ever more complex representations were supposed to be built, ever more accurately representing the different visual aspects of the environment. The final stage of the way objects were represented was by means of specifying their complete three-dimensional layout in an object-centered coordinate system. To render this concrete, think about the equations of a circle as the set of points equidistant from a given point. This gives a specification of the shape of the circle in terms of a fixed point. If one replaces the point by three oblique lines—or coordinates—at the center of the shape, one can specify a three-dimensional shape by providing the distance of each point on its contour from these coordinates.

Marr supposed the visual system worked, in its initial stages up to the 3D representation, independently from nonvisual knowledge about the world. Such knowledge, for example, concerns which categories seen objects belong to, such as which forms can be categorized as chairs or as furniture. The purely visual stages rely only on visual knowledge. Visual knowledge should be understood as the reason behind the computations that are carried out. For example, the form system will target and operate on spatially coherent gradients in light intensity across an image, because these often coincide with an object's contours that distinguish it from its background. General knowledge, such as concerning the general layout of chairs, is left out of the equation in the purely visual stage. Hence, Marr's theory is "bottom-up"—it doesn't implicate "high-level" conceptual knowledge in the initial stages of visual information processing. The knowledge that a chair normally has four legs doesn't shape the flow of computation in properly visual processing.

So, Marr's theory is about "pure vision" (Churchland et al., 1994). But how does this pure vision relate to what organisms see, to what they know, and how they act? Marr says, "Vision is a process that produces from images of the external world a description that is useful to the viewer and not cluttered with irrelevant information" (Marr, 1980, p. 31). But this raises a number of questions. How does the visual system or a neural process acquire the possibility to describe things? A person can describe what they see by saying things, but how does a neural process do this? Who is hearing or reading the description? Or, in the cognitivist framework, where is the "viewer" the description is useful to? How does the description lead to action? We stumble upon an old problem, known at least since the time of Descartes: that of the homunculus (see the Bibliographical Note for Descartes's formulation of it).

If one thinks of vision in terms of an inner image—a picture in the brain—doesn't this require an inner viewer of the image—a little person looking at the picture in the brain? But if one needs an inner viewer to explain seeing, doesn't the seeing of the inner person require yet another viewer, inside the inner viewer? And so on. One stumbles upon an infinite regress problem (see section 7.3.2 of chapter 7). Indeed, Ryle raised an infinite regress problem not only with respect to rule-following, but also with respect to seeing (Ryle, 1949a).

Cognitivists like Marr answer that the visual description is linked to action because of further computational processes. Purely visual information is combined with information about action, and as a result, actions can be planned and executed. This information processing or computation is thought of as creating and operating on representations. Computational operations are cognitive operations. They involve the following of rules. A computation, even a simple one as performing addition, is not just a physical process going on in the computational system; it is a cognitive operation because it entails the transformation of a representation in a rule-respecting way—in our example, the rules of addition. By describing basic computational operations allegedly underlying the cognitive activities of persons and organisms, cognitivism is a "cognition first" approach. So, in the end, Marr's computational theory of vision still does what Wittgenstein and Ryle criticized: It explains the cognitive activity in terms of underlying cognitive activity. For Wittgenstein and Ryle, the computational strategy for avoiding the homunculus by invoking computations consists of pushing the crucial problem a level down. Cognition at the deeper level would be just assumed, not explained. In chapters 10 and 11, we'll see how later thinkers have revived the opposition against the "cognition first" approach and have proposed ways of exploring cognition in different ways.

But first let us look at another example of "cognition first" in action, in the work of a very prominent cognitivist theorist: Noam Chomsky. In fact, with his critical review of Skinner's book *Verbal Behavior*—invariably characterized as "devastating" in cognitivist texts—he became in many respects the face of the so-called "cognitive revolution" in which behaviorism was replaced by the cognitivist approach (see chapter 4; also Gardner, 1985). In his early career he published a book titled *Syntactic Structures* (Chomsky, 1957). The work contained a technical analysis of the grammatical structure of natural language sentences. Chomsky posited that such grammatical structure could be studied "autonomously": without regard for, among others, the meaning of sentences or actual language use. He further argued that all natural languages have a complex *hierarchical* and *recursive* structure. A structure is hierarchical when base-level components combine to form conglomerates that themselves function as components of a higher-level structure. There can then be an order of base components and an order of conglomerates. The sentence "Lucy sees John" has just three components. "The girl with the blue eyes sees John" has eight base components. Yet the two sentences have the same structure: They are formed by a noun phrase, a verb, and another noun phrase. They have only three second-structure components. Further, hierarchical structures can become recursive. This happens when a component can have the same structure as a whole, like in "John sees the girl seeing John." Recursion can repeat, as in "The man seeing John sees the girl seeing John," and this could be further recursively expanded. Chomsky argued humans can understand and generate language only if they possess some internalized rule-system, matched to the recursive complexity of grammar.

Chomsky developed a particular view regarding language learning, namely, that language could be learned only if language learners already had some capacities for reasoning: for forming hypotheses and testing these against evidence. Chomsky claimed that this assumption was necessary because, in his view, children learning a language only get poor learning material. This "poverty of the stimulus" argument, as Chomsky (1980, p. 34) named it for language learning, is analogous to Mill's and Helmholtz's argument for the need for unconscious inferences to explain perception—an argument that also motivated cognitivist computational approaches to perception in the visual and other modalities. What's given to the visual system is supposed to be impoverished and in need of inferential or cognitive processing because it doesn't provide the information we get in perception. What's immediately available are merely blobs of color (as, e.g., in John Stuart Mill, see chapter 3) or ambiguous retinal images (as in Helmholtz—see chapter 3, or in Marr—see the current chapter), what's given in perception is a seen scene. Because of these differences, inference or computation has to step in to arrive at perception.

Chomsky's linguistic version of the poverty of the stimulus argument rests on the assumption that the language utterances that very young children are confronted with—what parents and peers say to them, and how they engage in proto-conversations—are insufficient to figure out what the correct grammar of the language is. That is, in Chomsky's view, the child is at pains to infer the correct structure of the language. The child can form hypotheses about what the structure is, but it doesn't get sufficient feedback to correctly choose between alternative hypotheses. Therefore, says Chomsky, the child must have a hypothesis-forming system that is already tuned toward the actual structure of languages: one that, from the start, leads to the approximately right hypothesis, which can then be further confirmed by the data. He terms such a hypothetical system a "language acquisition device," or LAD.

Though the technical elaboration of Chomsky's views on language have changed considerably in his subsequent publications, the basic ideas have remained stable. The learning and use of language is still deemed possible because of some recursive system hardwired in human brains. Such recursive system underlies not only language but also inference and planning. In a book written together with computer scientist Robert C. Berwick (Berwick & Chomsky, 2017), Chomsky describes the system as an "'internal' trait that boosted selective advantage internally, by means of better planning, inference, and the like" (p. 164). Moreover, the system is assumed to have arisen selectively, but not in a gradual way. Rather, one specific mutation is supposed to be at its origin, a single mutation that brought about sharp discontinuity of function (Berwick & Chomsky, 2017, see also Stuart-Kennedy & Terrace, 2017).

Chomsky thus supposes that there is (innate) grammar before there is actual grammar, an innate complex planning capacity before there is actual complex planning. Moreover, he claims that there is no gradualistic kind of explanation for these capacities. This opposes him to Darwin and, later, Skinner, who took such gradualism as a core feature of evolutionary explanations. A single chance mutation delivering a grammar or planning system would be highly suspect to them and not offering an evolutionary explanation at all.

Chomsky himself points at the similarities between his approach to language and cognition and the thinking of Descartes (Chomsky, 1966). Indeed, Chomsky's insistence on the need for assuming a special computational system at the basis of language is motivated by the belief that it is the only way to account for a human capacity that is unique to humans because of its general and generative nature.

"Cognition first" is exemplified in Chomsky's work in the picture of language he arrives at: a picture of language as relying on something with the essential features of language, residing in the brain. The language you'll speak is a result of the language you already have. But "cognition first" also animates Chomsky's argument for that picture. The poverty of the stimulus argument assumes that children that learn a language do so by forming hypotheses, by reasoning and inferring on the basis of input. But forming hypotheses, reasoning, and inferring are all cognitive operations. More than that, they are linguistic cognitive operations. Again, in Chomsky, the "cognition first" approach becomes a "language first" approach.

Do Chomsky's ideas about an innate grammar get support from the more recent trends in evolutionary thinking in which gradualism is being reconsidered (see section 4.7 of chapter 4)? Not quite. There is gradualism and then there is gradualism. The rethinking of evolution as something that itself evolves described in Lala et al. (2024) allows for the possibility of nongradual evolution. At the same time, the approach is empirical: The case for nongradualism is made for specific examples of nongradualist evolution, about which there is a great amount of knowledge about how the nongradual evolutionary steps could have been made possible by previous evolutionary steps—for example, in evolution traits can have become connected because they are results of a single developmental pathway. But then the pathways are known in biochemical detail (Lala et al., 2024, part II). Lala and colleagues give no free pass to help oneself to any nongradual step. Further, the observational material that they point to when considering the evolution of language is alien to what Chomsky had in mind. Instead of finding evidence for a "proto-grammar" mutation, they point at evidence for adaptations that facilitate social interaction and learning. Overall, they hold a complex picture of language as both a result and a driver of increased sociocultural evolution. They don't take grammar as we have it in our languages as evidence for a preexisting grammatical capacity that lies at the basis of our cognition. Rather, complex grammar emerges along the way out of changing ways of communicating that have a less complex grammar. Complex rules might be needed only when the vocabulary gets bigger. The most irregular verbs are the most used and probably oldest ones: like *having* and *being*. Once more verbs enter the mix, rules for dealing with them might appear so that people don't have to memorize each and every conjugation (Henrich 2016, chapter 13). Grammar, so conceived, is a result, not something at the basis.

7.3.3 Cognitive Science Evolving

The core ingredients of classical cognitive science—internalism, computationalism, and representationalism—naturally supported the feasibility of cognitive machines, or artificial intelligence. Cognitivist ideas about the primordial role of representations provided fertile ground for confidence in the possibility of creating artificially intelligent computers. If cognition is basically a matter of processing information, or of transforming representations into other representations, it could equally well be found in computers as in human beings. The same information or the same representational meaning can be carried by different means. "*Il pleut*" in French, "it rains," and "IT RAINS" all carry the same information, or represent the same fact. Moreover, the sentences could be drawn in the sand of the shore, accidentally formed by clouds in the sky, or tattooed on one's elbow in purple and green. Despite having very many different substrates and formats, these remain the same representations, carrying the same information. Moreover, if brains are processing information by performing computational operations on representations, the substrate—brainy tissue—in which the computations happen doesn't matter either. Computation as conceived by Turing and the later cognitivist theorists, even if derived from human activity, was by definition mechanical and, hence, machine-executable. Finally, internalism warranted that only a part of the body, corresponding to the human brain, was essential for cognition to take place.

So, from the beginnings of cognitive science, there was great optimism about the feasibility of artificially intelligent machines. If machines were fed the right representations and applied the right computations on them, they could perform, or at least simulate, humanlike cognitive acts. For example, if a machine would receive something informationally equivalent to our retinal images, and it would compute the appropriate visual representations, it would have artificial vision. It would, just as the human visual system was supposed to do for Marr, construct a representation of the outside world that made explicit the visually crucial features such as three-dimensional form and spatial position. Marr termed such representations "symbolic," because the computational activity in the visual system was aimed at a descriptive end product.

Also, in many other branches of cognitive science, a symbolic approach was taken. For example, in linguistics, the goal of the cognitive system responsible for understanding meaning was the construction of a semantic representation, understood as a symbolic description of meaning (Shanon, 1988). In general, knowledge was taken to be produced and stored in symbolic format: as in some way analogous to public symbolic systems like natural languages or mathematics.

This overall approach led to a belief that one could build an artificially intelligent system with cognitive competence in a certain domain by uploading the domain-specific knowledge and programs to computationally manipulate this knowledge. In this vein, an "expert system" called MYCIN was created at Stanford University in California that could diagnose blood infections. MYCIN stored in memory a large number of domain-specific rules, such as: "If the sugar value in the blood level is higher than 100, this may indicate a preliminary stage of diabetes." By combining such rules, together with specific data fed to it, the system could arrive at a diagnosis of a given case (Buchanan & Shortliffe, 1984). Ultimately, a system like MYCIN performed at the level of a general practitioner and comparable to a doctor specialized in blood disorders. The symbolic approach also led to machines that were able to play games such as chess at a level of, and arguably sometimes above, the best human players—interestingly, the basic ideas for programming a computer to do so were first published by Claude Shannon (1950).

However, it soon became clear that the symbolic representational approach had limits. It worked quite well in domains that are very regular and structured by rules. Games such as chess are a paragon of that. But many other domains that humans cognitively operate with ease and intelligence in, are less regular and structured. Consider what people do when they go to a restaurant. Often, what they do conforms to a general restaurant script: seeking a table, ordering with a waiter or nowadays with one's phone, getting the food, using the cutlery to eat, and so on. But how do people take and operate their cutlery? That will depend on what's on the plate. If one applies too much force on a hard piece of food, it might be launched as a projectile. If too little, it might just remain where it is. Fork and knife might have to be used to reposition a bigger piece. A multitude of possibilities opens up, yet humans can deal with them easily. Further, people will know what to do if the waiter sneezes into the food being served, if they see a mouse finding its way around the table, and even if a giant kangaroo appears at the outside of the window. None of these circumstances can be foreseen. It would be impossible to capture in a symbolic knowledge structure like a set of rules, all the possible contingencies to which humans—and animals, for that matter—can react appropriately. This flexibility of human cognition was noted and contrasted with that of computers by philosopher Hubert Dreyfus (1929–2017). He developed a critique of artificial intelligence in *What Computers Can't Do* (Dreyfus, 1972). This book, later updated and reissued as *What Computers Still Can't Do* (Dreyfus, 1992), became a classic.

More generally, Dreyfus criticized representational and computational views as they were endorsed in early cognitivism. Relying on the French philosopher Merleau-Ponty (1908–1961), but also Wittgenstein, Ryle, and Heidegger, Hubert Dreyfus developed a critique of the overall cognitivist conception of cognition, especially as it inspired research in the early phases of the novel discipline of artificial intelligence. Dreyfus criticized conceiving of cognition and cognitive processes in terms of language-like systems of symbols, then operated upon by programs, that is, of systems of rules. The crucial mistake that researchers in artificial intelligence and cognitive science committed, according to Dreyfus, was to confuse skills with some form of theoretical knowledge. Intelligence often relies on skills: being able to do something. Often skills are learned, as in the case of cycling or juggling, or gradually acquired over a period of time. For example, we acquired the ability to write, read, and do mathematics during our childhood step by step. Skills can be contrasted with certain forms of theoretical knowledge that are represented as sequences of rules. That kind of theoretical knowledge can be found, for example, in your smartphone instruction manual or in a drug's package insert. A defining feature of skills, Dreyfus argues, is that they can be described only to a very limited extent with explicit rules. How exactly do you tie your shoelaces?

Moreover, skills adapt smoothly to all kinds of circumstances, even if these have never before been encountered. A good juggler will be able to perform their tricks despite wearing gloves and, possibly with some effort or practice, even when they put on a gorilla suit. Our skills have a flexibility and adaptivity that rule-guided theoretical knowledge does not have. If you are going to start modeling skills on explicit rules, you get rigid and, therefore, fragile pseudo-skills. Explicit rules, Dreyfus argues, can and often do play a role in the acquisition or learning of skills. You can guide the learning process by providing guidelines for the actions to be practiced. However, the skill is not acquired by memorizing those rules, but by trying actions many thousands of times and gradually getting better at them.

The kinds of issues that troubled Dreyfus became known as problems of "brittleness" (Clark, 2014, chapter 2; Winograd & Flores, 1986). They refer to a lack of flexibility or adaptability of cognitive systems: not being able to cope if a situation even slightly differs from what the system is used to dealing with. The problem is not specific for symbolic AI—it appears in cognitive psychology whenever driven by the same theoretical framework.

One way cognitive scientists tried to deal with brittleness was by developing what came to be known as artificial neural networks. Though some pioneers have experimented since the 1940s and 1950s with networks of artificial units resembling brain cells to some extent, it was only forty to fifty years later, with the so-called "connectionist" movement, that building artificial neural networks became a widely adopted method for studying cognition and exploring artificial intelligence.

The natural starting point for neural network research is to produce artificial brain cells. Brain cells or neurons are the main components of the brain. They consist of a cell body and connective channels. Through these connections, cells can affect each other's level of activity. Nerve cells can be on or off, or something in between. If they have reached a certain level of activity or excitation, they will "fire" by activating outgoing connective channels. In this way, the channels between the neurons—the axons and dendrites—allow one cell to control the activation level of other cells.

Frank Rosenblatt (1928–1971) started the Perceptron project in 1958, an early attempt to design a neurally inspired machine. Rosenblatt wanted to create an artificial neural network—a series of interconnected artificial brain cells—that could, based on camera images, decide which category the image belonged to. Categorization is a fundamental requirement for cognition: To cognitively or intelligently respond to the environment, one has to treat different types of things differently. You only want to pick edible mushrooms, for example, and leave the poisonous ones where they are. The perceptron system would get as input an image—a large number of pixels, each in its own place—and it has to respond differently to dogs than to cats. If the network systematically shows a specific cat response to cat inputs, and a specific dog response to dog inputs, the network classifies dogs and cats. To accomplish this, Rosenblatt built a machine that consisted of two layers of neurons: a layer that received input, and a response layer that indicated whether the input concerned a dog or a cat.

A two-layer neural network, however, can perform only a limited number of classifications (Minsky & Papert, 1969). As a result of such limitations, perceptron research was abandoned. It was not until the 1980s that the thread was picked up again and research into artificial neural networks regained momentum. *Connectionism* emerged—a term referring to the connections between neurons. Unlike Rosenblatt, connectionists did not aim to build an artificial brain consisting of artificial nerve cells. Instead, they designed simulations of neural networks that ran on "ordinary" computers. Working with simulations, connectionists explore more complex networks than Rosenblatt did, that is, networks with one or more layers of neurons between the input and output layers.

A typical connectionist neural network consists of several layers of artificial neurons that have connections with each other (Clark, 2014, chapter 4). Those connections can be weaker or stronger, and these strengths can be adjusted. Just as with real brain cells, the simulated cells affect each other. That is, if a simulated cell is active, it will transmit activity to the cells with which it is connected—to a lesser or greater extent, depending on the strength of the connection. Based on the incoming activity, the receiving artificial neuron will itself become active to a certain degree and transmit activity itself. Very importantly, the strength of the connections between simulated neurons can change. Here lies the key to "training" the neural network.

Suppose a researcher wants to build a network to categorize what's shown in a set of pictures. To do so, they will choose a network with an input layer, a number of (hidden) intermediate layers, and an output layer. The input layer will contain cells that are activated by parts of the picture. The lighter the corresponding part of the picture will be, the more active the input layer cell will be. The activity of the output layer will indicate which category what's pictured falls into. In the simplest case, the output layer could consist of two neurons to allow for a simple perceptual discrimination. If the activity of neuron *a* is then greater than the activity of neuron *b*, this would indicate the picture is taken by the network to show something belonging to category A, and if the reverse is the case, it would be in category B.

Then the training process begins. At the start of the process, the value of all the connections between cells is randomly determined. The researcher shows the network a picture and then finds out which activation in the output layer follows. If the activation corresponds to a wrong response, they make changes to the connections—particularly to the connections that contributed most to the wrong answer. In the case of a correct answer, they do the opposite: strengthen the connections that contributed most to the correct answer. This process is called "error backpropagation" (Clark, 2014, chapter 4). It has been established that after many (hundreds of thousands) of iterations of this process, the network will be able to provide the correct answer. Moreover, the work of the researcher can be automated. For example, the values of the connections can be adjusted by an algorithm. Connectionists designed neural networks that could be successfully trained in, among many other things, identifying faces, converting written into spoken language, and recognizing grammatical patterns in language. Remember that the name "neural networks" is somewhat misleading here, because almost all of the research on connectionist neural networks was conducted not on networks of artificial neurons but on simulations of such networks on digital computers. What neural networks essentially do is become sensitive to overall patterns that are characteristic for certain categories (Churchland, 1989, 1995). Roses are recognized by the network as roses and not as another kind of flower because the network is sensitive to a very large number of features that tend to be present in some combination in roses and not in other flowers.

Neural networks gave some theorists hope for advancing artificial intelligence, because these networks had special features. For example, artificial neural networks can correctly categorize a picture of a rose if the picture is only partially presented. It can respond to the overall "rose pattern," even if parts of it are missing. To give another example, a neural network can be made to lose a number of its nodes and connections and still function relatively well—this feature is called "graceful degradation." To some extent, features such as these imply that connectionist networks are not as "brittle" as standardly programmed computers are. Moreover, to some extent, connectionist networks acquire their "cognitive" capacities in a way comparable to how people acquire some of their cognitive abilities: by being trained, with recurrent feedback, on specific cases and examples. Remember that Dreyfus pointed to this way of achieving competence in his critique of the nonconnectionist approach to artificial intelligence.

Though in many ways the connectionist approach to cognition differed from other computationalist approaches, many connectionist theorists retained the standard tenets, or at least the standard vocabulary, of the overall cognitivist approach (Clark, 2014, chapter 4). They saw connectionist networks as engaged in computing, to categorize input. Connectionist networks were said to encode knowledge in the strength of their connections. A pattern of activation of a collection of nodes of its output or intermediate layers was taken to form a representation of a category.

Yet some cognitive theorists were of the opinion that connectionist networks missed essential features of computational systems closer to Turing machines. Jerry Fodor and Zenon Pylyshyn, for example, argued that neural networks just associate inputs with outputs. In their opinion, these networks are not able to create and operate on complex hierarchical, recursive structures, such as in arithmetical formula or when, in a natural language, a sentence contains another sentence—as in "She regrets he was late." On Fodor and Pylyshyn's view, connectionist networks have no means to represent and operate on symbolic formula, for example, by replacing one symbol with another or changing the order of constituents. In Fodor and Pylyshyn's view, such kind of symbolic cognitive system must be assumed to be behind cognition—just as a similar inner language device is supposed to be behind language by Chomsky. Because connectionist networks lack what Fodor and Pylyshyn call a "classical cognitive architecture," they would be unable to get a grip on the structural similarity between "The cat chases the dog" and "The dog chases the cat." Perhaps connectionist networks, say Fodor and Pylyshyn, could learn to treat these inputs as belonging to one category, but they would never do so because of a sensitivity to its inner structure (Fodor & Pylyshyn, 1988).

Connectionist researchers responded to this criticism in many ways (Clark, 2014, chapter 2). Some defenders argued that connectionist networks, when analyzed well, could be shown to have the right sort of sensitivity to structural features of input (Churchland, 1989), or could be made to have such sensitivity by extra design features—such as being architected to feed its prior output back into itself upon receiving new input (on these recurrent networks, also see Clark, 2014, chapter 4).

Yet, for all intents and purposes, connectionist research eventually reached a plateau, after a surge in the 1980s and 1990s. Only in the twenty-first century, a connectionist-like approach re-entered the main stage. In the intervening period the computing power of computers had increased dramatically, and a new generation of programs has emerged that are inspired by the original connectionist neural networks. The new approach became technically known as "deep convolutional neural networks" (DCNNs), in wider culture as "deep learning." DCNNs comprise far more layers than the earlier connectionist networks (Buckner, 2019, 2024). The units of deep learning networks are not, or even less so than the artificial neurons of the connectionist networks, meant to be biologically realistic analogs of brain cells. They are designed to optimize the classification capacities of DCNNs. And this goal has, to a surprising extent, been achieved. DCNNs have quickly become standard technology. Common applications are automatic recognition of faces in smartphones, programs for language translation, or large language model-based systems like chatbots. All such systems are, at base, classifiers or pattern recognizers—like the earlier neural networks, but with much more power. They achieve their functionality by finding patterns in sometimes massive amounts of data and by producing outputs that cohere with such patterns—this is how a chatbox delivers its answers: by finding answers that are "lingering" in text fragments available on the internet.

Interestingly, DCNNs, like connectionist networks, but at a much larger scale, employ the "exposition to a multitude of examples" strategy that Hubert Dreyfus saw as a key to human intelligence. Do DCNNs defy the criticism that Dreyfus addressed to the AI of his time? Do DCNNs show that genuine humanlike artificial intelligence is within grasp, or perhaps already achieved?

Some old-school cognitivists such as Noam Chomsky resolutely answer "No," based on a belief in a fundamental difference between associationist and classical computational cognitive processes (Buckner, 2024). The overall concern is that a large language model–based system does nothing more than extract patterns in data that were originally produced by humans. Humans, on the contrary, have original generative capacities.

But one need not endorse the existence of such a Chomskyan distinction between cognitive processes to be nuanced about the relation between the kind of intelligence shown by DCNNs and organic cognition. Contra the Chomsky-like generative line, defenders of DCNNs might argue that humans, too, in their writing, speaking, and thinking, extrapolate from patterns they are exposed to. One can admit that—or at least to some extent—and still not grant DCNN systems the kind of intelligence that humans and other organisms have. Proponents of embodied or enactive cognition—see chapters 10 and 11—have pointed, instead of to some inner difference in cognitive processing, to the fact that they are disembodied (Chemero, 2023). They have no bodies, no metabolic needs, and no ways of satisfying them. They are not born, raised by parents or caretakers, nor are they in any credible sense engaged in social interactions. They don't perceive, or move to better perceive, and neither can they be emotionally affected by what they see, hear, or touch. Some strands in embodied thinking of cognition take humanlike cognition to be essentially linked to being a biological, living body and will see DCNNs as a low-dimensional projection of some specific human capacities, rather than as those same capacities "realized" without the embodied and social aspects. In fact, they claim that it is impossible to "realize" those capacities while leaving out the socially embedded body.

Bibliographical Note for Chapter 7

Andy Clark's *Mindware: An Introduction to the Philosophy of Cognitive Science* contains much excellent discussion related to the rise and further development of cognitive science and its philosophy (Clark, 2014). It is relevant to all of the subsequent chapters of this book.

I have omitted many recent developments in cognitive science. Some of these, such as embodied and embedded cognition, will be dealt with further on. But I have said nothing here on the important and highly influential trend called "predictive processing." Predictive processing accounts aim to provide a comprehensive view of cognition as probabilistic inference. Its origin is often traced to Helmholtz's notion of unconscious inference. Roughly, predictive processing views cognition as the process of inferring what is probably the case, given the current stimulation. This process of recovering the probable causes of stimulation is based on a sense of, or model of, the probabilistic structure of the environment. The cognitive system receives cues in the form of sensory stimulation and then forms hypotheses about what's likely the cause of that stimulation. Hypothesis formation is guided by knowledge of statistical regularities in the environment. If noise on the roof has in the past been almost always caused by rain, the hypothesis will be formed, upon receiving noisy roof stimulation, that it is raining. As I am sketching it, predictive processing sounds very cognitivist and intellectualist—modeling cognition in general on thinking. Some versions of the predictive processing approach have tried to shed this intellectualism. For an intellectualist approach to predictive processing, see Hohwy (2013); for a less intellectualist version, see Clark (2015).

Descartes formulated his version of the homunculus criticism in a passage in his *Optics*, where he wrote:

> The images of objects are not only formed in this way at the back of the eye but also pass beyond into the brain. . . . Now when this picture thus passes to the inside of our head, it still bears some resemblance to the objects from which it proceeds. As I have amply shown already, however, we must not think that it is by means of this resemblance that a picture causes our sensory perception of the objects—as if there were yet other eyes within our brain with which we could perceive it. (ATVI 128–130; CSM1 167)

8 Mental States and Functionalism

How does the philosophical problem about mental processes and states and about behaviorism arise? The first step is the one that altogether escapes notice. We talk of processes and states and leave their nature undecided. Sometime perhaps we shall know more about them—we think. But that is just what commits us to a particular way of looking at the matter. For we have a definite concept of what it means to learn to know a process better. (The decisive movement in the conjuring trick has been made, and it was the very one that we thought quite innocent.)
—Ludwig Wittgenstein (1953, § 308)

8.1 Philosophy and Cognitive Science

From the beginning, cognitive science included philosophy (Gardner, 1985). In fact, cognitive science was welcomed by many philosophers thinking about the mind as allowing to reconnect with a philosophical tradition that was criticized substantially on the one hand by Carnap-style positivism and, on the other, by people like Wittgenstein, Anscombe, and Ryle. In the days of the cognitive revolution, these views were increasingly seen as overly restrictive, as imposing constraints on philosophy and science that were too stringent. The views mentioned became portrayed as philosophical varieties of behaviorism. The cognitive scientific revolution that was very much directed against behaviorism in psychology was paralleled by a rejection by philosophers of the view that came to be called "philosophical behaviorism." Philosophical behaviorists were taken to deny the existence of inner mental processes and inner mental states. Abandoning philosophical behaviorism then meant re-endorsing the inner in the philosophy of mind. The philosophers that joined the cognitive revolution admitted that the identity theorists had gone some way in reviving the inner, but it was believed that a philosophy allied to cognitive science could do justice to the inner in a much more profound and fruitful way. The computational view of mind was held to give scientific credibility to a picture of the mind as in important ways inner. In many ways, mainstream philosophy of mind as it arose alongside the cognitive revolution reconnected with the philosophical tradition that led to Locke and Hume's views of the mind. In previous chapters (especially 3), we have seen how this tradition has much influenced subsequent philosophy, as well as scientific psychology.

So, just like cognitive scientists identified themselves by contrasting their views with scientific behaviorism, like-minded philosophers drew on contrasts with what came to be called "logical behaviorism," "analytical behaviorism," or simply "philosophical behaviorism."

To be a logical or analytical behaviorist is to believe that the meanings of mental concepts can be given in terms that refer only to outward behavior. It is a thesis about the definitions of mental terms, hence, the "logical" or "analytical" in its name. Positivist-minded philosophers, notably Hempel and Carnap, were taken to have been logical behaviorists, as well as Ryle, and perhaps to a somewhat lesser extent, Wittgenstein and Anscombe.

Analytical behaviorism, as it is defined and refuted in many treatments, both in papers and in textbooks, consists of a denial that mental terms refer to inner states, episodes, or events, together with the assertion that mental terms can be defined—analytically, that is, with preservation of meaning—in terms that refer only to behavior. Thus, William Lycan (Lycan, 1990, p. 4) characterizes "analytical behaviorism" as:

> the claim that mental state ascriptions simply *mean* things about behavioral responses to environmental impingements. Thus "Edmund is in pain" means, not something about Edmund's putative inner life or any episode taking place within Edmund, but that Edmund either is actually behaving in a wincing-and-groaning way or is disposed to behave so (in that he would so behave were something not keeping him from so doing). "Edmund believes that broccoli will kill him means just that if asked, Edmund will assent to that proposition, and if confronted by broccoli, Edmund will shun it, and so forth.

Analytical behaviorism, thus construed, aims to reduce mentality to movement. Further, the movements mentality is reduced to are understood as *mere* movements: displacements of bodies and limbs in space. No allowance is made for the context a movement takes place in. Movements are not actions.

Many articles and textbooks contain refutations of this analytical behaviorism. A classical source is Putnam (1960). The most recurring arguments—present in Putnam's text but also in countless others—are twofold. First, it is pointed out that one could have the movements without the mentality and mentality without movement. For example, someone, say an actor, can feign all the movements of pain but have no pain at all. This, then, is taken to show that one cannot reductively identify the mental phenomenon of pain with pain movements. Therefore, analytical behaviorism is wrong. By uniquely focusing on movement, it is missing what's most important for the mental: its inner aspects. If the actor example illustrates how one can have movement without mentality, another famous example, Putnam's "super Spartans," illustrates the converse: that one can have mental phenomena without movement (Putnam, 1960). A community of specially trained super Spartans could suppress all pain behavior, even when confronted with severe pain. They have inner pain, despite showing no outer pain movement. The example shows, so the argument runs, that there is more to mentality than just movement, and therefore analytical behaviorism is mistaken. The second major objection to analytical behaviorism is that definitions of mental terms in terms of movements will never be complete. Which movements anyone having beliefs about food will show will depend on what this person further believes, hopes, knows, and so on. If a hungry person holds the belief that the food offered to them is poisoned, they will not eat it. But then movement, so the argument goes, depends on the mental, and not the other way around.

A forceful defense and elaboration of a postbehaviorist repositioning of the philosophy of mind can be found in an influential book by Australian philosopher David Armstrong, A *Materialist Theory of the Mind* (Armstrong, 1968). In this book, Armstrong proposes that the mental should be understood as fundamentally involving causal states. The mental concerns mental states, and mental states are causal states. A causal state is a state that is defined as the effect of certain causes and the cause of certain effects. How can the idea of causal states be fruitful to analyze the mind? Consider the sensation of green. In the causal state analysis, the sensation of green is a causal state that is the effect of certain stimuli and the cause of certain behaviors. Stimuli can consist in being presented with grass, some kinds of apples, and so on. Behaviors encompass, for example, sorting something with other green things, or saying "This one is green." Crucially, Armstrong insists that the causal connections of a mental state can involve mental states. Mental states not only causally connect to stimuli and to behavior but also to other mental states. Thus, having a purpose is analyzed by Armstrong as a mental state that stands in causal relations not only to specific worldly events and bodily activities but also to certain beliefs. Such an analysis is thought to be needed to accommodate the fact that one will act on one's purposes only in the right external conditions and given certain beliefs. Consider again a hungry person. They are in the purposive state of wanting to eat. But a plate of food will not move this person's purposive state to affect the appropriate muscles if the person is in the belief-state that the food is poisoned. Armstrong followed the identity theorists in claiming that mental causal states should be identified with states of the brain or central nervous system. Hence, he termed his approach "the central-state theory" (Armstrong, 1968, chapter 6).

When Armstrong's plea for a causal understanding of the mental appeared, another idea had been launched that would be combined, by Armstrong himself and by many others, with the causal understanding: the idea that in our daily interactions with other human beings we rely on a theory of each other. In our daily lives we all use a specific vocabulary to describe each other: We say we feel, perceive, believe, want, fear, and so on. That we have a special vocabulary for talking about minds had not escaped the attention of philosophers. As we have seen in chapter 6, Wittgenstein, Anscombe, and Ryle were very much occupied by explicating how we use mental terms. But a new idea appeared on the scene: that this mental vocabulary was part of something akin to a scientific theory, like physics or chemistry. Our mental vocabulary now was supposed to be part of what came to be called "folk psychology." And, we were supposed to rely on our tacit knowledge of this theory to predict and explain each other's behavior in our daily lives.

A very influential source for this idea is the text, "Empiricism and the Philosophy of Mind" by American philosopher Wilfrid Sellars (1912–1989; Sellars, 1956). In this long and many-facetted essay, Sellars introduces a fictive account of how the vocabulary for mental phenomena often thought of as "inner" could have originated. "Inner" mental phenomena are those like imagining or thinking. Compared to seeing, imagining is not necessarily tightly coupled to the environment. One can only see a tree if it's in one's view, but one can imagine one while in a rocket nearing the moon. One cannot say, "You could have shown some more respect" without speaking, but one can think it while holding one's lips still. Sellars imagines a time when people had a vocabulary only for seeing, speaking, and other "outer" phenomena. But then a person, called "mythical Jones" by Sellars, comes around. Jones makes the brilliant move of expanding the language. He introduces that besides these observable events of seeing and speaking, invisible episodes that are *like* seeing and *like* speaking occur in his fellow human beings. Jones then refers to these episodes to explain what other people do not only in terms of what they have seen and said, but also in terms of what is "seen" (visualized or imagined) or "said" (thought or believed) in these private episodes.

The Jonesian step thus comes down to positing in-itself-unobserved hypothetical somethings that help in accounting for observed behavior. This seems to be what scientists do all the time: To account for the observed patterns in the behavior of objects, they posit atoms, forces, and ultimately elementary particles like fermions or bosons. These are items that are themselves not observed, but plausible causes of what is overtly visible. Of course, what is now still unobserved can later become observed. For example, with increasing advancements in imaging technology, what was unobservable for early physicists is now observable. On the Sellarsian picture, insofar as we understand each other in terms of "internal states" like imaginings or beliefs, we engage in something very similar to what scientists do. We apply a psychological theory for purposes of explaining and predicting. We all share the theory—which became termed "folk psychology"—without having had to go to college to obtain a degree in it.

Armstrong himself was attracted to Sellars's idea that our mental vocabulary belonged to a kind of theory (Armstrong, 1968, chapter 5). But it was his Australasian compatriot David Lewis (1941–2001) who forged a very elaborate and detailed combination of the idea of the mental as a matter of causal states with the idea of folk psychology (Lewis, 1966, 1970, 1972). He asserted that the way we already speak about the mind is along the lines of Armstrong's causal theory. When we speak of people being in pain and having beliefs and intending things, we are applying concepts defined in terms of causal profiles. "Pain" means something along the lines of: the state that plays the role of being caused by harmful objects, that is connected to states of intending to seek relief, and to behaviors like screaming or shouting "ouch." Lewis agreed we use mental vocabulary to explain and predict behavior. When we say that a person calls the doctor because they intend to have an examination of the pain they believe to be due to a stomach ulcer, we explain what the person does in terms of inner episodes, behavior, and stimuli, and the ways these are linked. Lewis thought a careful analysis of our common mental vocabulary could be made that showed exactly which causal profiles were specific for each and every mental term. Such causal analysis would also allow us to see that the mental was not unique: Mental phenomena are just those phenomena that we use a special theoretical vocabulary for.

8.2 Functionalism and Multiple Realization

By this time, it had become common to speak of all mental phenomena in terms of states. The notion "mental state" became a ubiquitously used umbrella term to cover all things mental, from pain, emotion, to belief. Armstrong was committed to the "state" language and termed his own approach "the central-state theory." Yet Armstrong's central-state theory was still very much brain-oriented. "Central states" were meant to be states of the central nervous system. Thus, Armstrong retained a tight connection to the identity theory.

That was now about to change. Philosophers would soon distance themselves from the classical mind/brain identity theory. With the new developments in thinking about information, computation, and cognition in the background, the causal theory was lifted from the neurophysiological ground it was still rooted in with Armstrong. Several of the developments discussed in chapter 7 invited the moves that were to be made. For one thing, the "state" language was not necessarily connected to the nervous system. In particular, "state" talk was very central in the analysis of Turing machines. So were two other notions that entered the mainstage, those of "input" and "output." It was easy enough to link "stimulation" to "input" and "movement" to "output." It became appealing to reformulate the causal analysis of mental states offered in terms of "stimuli," "states," and "movements" by Armstrong, in terms of "inputs," "states," and "outputs."

The central state theory then becomes the functionalist theory of the mind. Mental phenomena are taken to be functional states defined by their relations to inputs, outputs, and other mental states. The mental state pain is the functional state one is in when something damages the body (relation to input), one goes into the mental/functional state of willing to move away from the damaging factor (relation to other mental states), and one cries out "ouch" (relation to outputs). Each distinct mental state is supposed to have its own functional profile.

This functional analysis might just seem like a causal state analysis with a somewhat new vocabulary of "inputs" and "outputs." But functionalism came with an additional claim: that of the "multiple realizability" of the mind. To claim that a mental phenomenon is multiply realizable is to hold that it can occur not only in people who have organic brains and bodies but also in systems made of completely different material. An alien creature could have pain, calculate a sum, or think, despite having a totally different physical makeup. And so could a robot.

For many, the role of computing and information processing lent plausibility to the multiple realizability of mind. Computers compute and process information, yet they have a different physical composition than human brains and bodies. All that matters is the right patterns of computation, the specific ways in which information is processed, or how inputs are transformed to yield outputs. Think about a calculating machine. To be a calculator, a device must handle its inputs to generate certain expected outputs. When given 512+813, it should yield 1,325; when given 11x11, it should return 121. What form the input, processing, and output takes doesn't matter. Input, outputs, and processing steps might consist in changing currents, pressures, or whatever. As long as certain overall patterns reliably occur, the system processes the same information. For information and its processing, it's the patterns that matter, just like, most of the times, the meaning of a sentence doesn't depend on which typeface or color it is written in.

Many functionalists endorsed the idea that at least some mental phenomena were multiply realizable and hence could occur not only in organic brains and bodies but also in systems made of completely different material. Functionalists saw multiple realization as marking a significant difference between functionalism and the identity theory. To make this case, they invoked a distinction between *types* and *tokens* (Fodor, 1974). Types are general. Tokens are particular things. "Dog" refers to a type, and "Fido" refers to a particular dog. The particular dog Fido is a token of the type *dog*. Philosophers also say that the token Fido realizes or instantiates the type *dog*. Now consider the type *pet*. The type *pet* has different subtypes. The types *goldfish*, *cat*, and *dog* are all types themselves that belong to the broader type *pet*. A token pet can belong to any of those subtypes. One token can realize the type *pet* by realizing the type *goldfish*; another token can realize the type *pet* by realizing the type *cat*. The type *pet* then can be said to be multiply realizable. Two token realizations of *pet* can realize any of the different subtypes of *pet*. There can be profound (sub)type differences between what belongs to or realizes the same overall type—as exemplified by our goldfish and our cat.

Now consider mental phenomena, conceived of as functional states. If pain is a functional state and is multiply realizable, the functional state pain can be realized not only in biological bodies but also in silicon bodies, alien bodies of unknown makeup, and so on. The mental type *pain*, if it is identical to a multiply realizable functional type, is not identical to a biological type, such as *pain brain state*. To identify the mental type *pain* with the type *brain pain state* would be to deny that there could be token pain realizers—those of aliens or robots, for example—of nonorganic, brainy or bodily types. It would be to commit an error of chauvinism, of unduly restricting the mental to the organic. Hence the functionalists argued that the identity theory was in error because it "type-identified" mental types with brain state types.

The correct move, according to functionalism, was to identify types of mental states with types of functional states. For functionalists, one could retain only a weaker mental/physical identity claim, namely, that every token mental state was identical to some physical state. In humans, and possibly other animals, the physical state was a token brain or body state. But in aliens, the token mental state would be identical to an alien brain or body state. So, in the end, many functionalists assessed the identity theory as being right about identities, but as messing up the distinction between types and tokens. Functionalists endorsed a token-identity theory for mental states but rejected a type-identity theory.

8.3 Varieties of Functionalism

Functionalism was put on the menu in several papers by Hilary Putnam (1926–2016) and appeared initially in a format that was tied closely to Turing machines (Putnam, 1965). The tape of a Turing machine was compared to the senses, and sensory inputs to the symbols on a tape. Movement was considered to be modeled by what the machines write on the tape. "Mental states," for example, pains or beliefs, were identified with machine states. It was insisted that these states were *internal*, so that what behaviorism allegedly left out of consideration was covered. Indeed, one of the reasons this "machine state functionalism" was taken to form an improvement over behaviorism was that it no longer considered behavior as a fixed response to a fixed input, but rather as something that depended both on input and on the current state.

Machine state functionalism also explicitly endorsed the multiple realization of mental states. For a system to have mental states, all that mattered were its symbols and the relations between its states. The physical details of how a symbol is written down, produced, or shown often don't matter. Further, it also does not matter what substance the machine is built from. Any machine with the same number of different states, which have the same causal relations with respect to each other, will function in the same way. At some later point Hilary Putnam expressed this hyperbolically as: "We could have been made of Swiss cheese and it wouldn't matter" (Putnam, 1975).

Nevertheless, it was by no means necessary for functionalists to adopt the framework of machines that were structurally or architecturally related to Turing machines. One influential way to develop functionalism differently was offered by Jerry A. Fodor, who proposed and defended "the computational theory of mind." Fodor's theory started with a certain view of folk psychology. According to it, the core concepts of folk psychology are "propositional attitudes." The term "propositional attitudes" was forged by Bertrand Russell in his 1919 paper "On Propositions: What They Are and How They Mean" (Russell, 1919). In Russell's treatment, propositional attitudes consist of a proposition, which should be understood, roughly, as the meaning of a descriptive sentence, plus an attitude that a person can take to that proposition. Thus, to the English and French sentences "It rains" and "*Il pleut*" corresponds the proposition that it rains. By believing, hoping, wishing, or fearing that it rains, a person takes one of multiple possible attitudes toward that proposition.

The term "propositional attitude" became widely used to talk about very common ways we have of describing each other. Saying that a person believes that it is going to rain, hopes that the sun will shine, or fears that lighting will strike are all taken to ascribe propositional attitudes to persons. There is an attitude component: that of believing, hoping, and fearing, and a propositional component: "that it is going to rain," "that the sun will shine," or "that lightning will strike."

But what exactly is the propositional component? It definitely consists of a phrase: "that it is going to rain," "that the sun will shine" or "that lightning will strike." But does the phrase just offer a way to describe a fact that the subject of the attitude has an attitude to? This seems plausible in cases like "The dogs hears a noise and believes her food is about to be served." Or does the sentence also serve to characterize the way the subject themself pictures the fact that is described by the sentence? This might seem plausible in cases like "He gave her the wrong present because he mistakenly believed *The Lion, The Witch and the Wardrobe* was written by Tolkien." It might be tempting to construe the propositional attitude ascription in the second example as committed to the assumption that the subject not only relates to a fact that is ascribed by the ascriber of the propositional attitude, but to a fact that is in some way pictured, characterized of described by the subject themself. Roughly, in the second case, propositional attitudes involve ascribing an attitude, not or not only to a fact, but an attitude to some kind of representation, held by the subject, of a fact. This way of viewing propositional attitudes has become widely shared as an approach to *all* propositional attitudes. Jerry Fodor has been one of its most impactful promotors .

Like other contemporary philosophers, Fodor held that propositional attitudes played an important role in folk psychology (Fodor, 1983). Folk psychology contained generalizations formulated in terms of schemas relating propositional attitudes such as:

If person P wants Y and believes that if X then Y, then, normally, P will want X.

By relying on such general schemas and filling them in with names for persons and sentences for propositions, people can predict and explain each other's behavior. For example, one could explain that Kim takes her umbrella, because she wants to go out, wants to stay dry, believes that it is raining, and believes that an umbrella will keep her dry.

Because Fodor endorsed the view that mental states and, hence, beliefs, desires, and all other propositional states were causal states, he claimed that the schemas of folk psychology, which he often termed "propositional attitude psychology," worked, precisely *because* propositional attitudes are real and causal states. The schemas track regular causal patterns between types of states. Kim's states of wanting to go out, wanting to stay dry, believing that it is raining and believing that an umbrella would keep her dry are states of her brain. If one had the technical means for it, one could observe them under her skull.

All of this stays firmly within the functionalist outlook. Kim's propositional attitude states are tokens of functional state types that are realized in the brain. Moreover, Fodor termed his view the "computational theory of mind", because what corresponded to the propositions in the propositional attitudes were taken to be symbolic or representational structures within a computational system (such as a Turing machine and, presumably, the brain). These structures were then operated upon in the system, or computational operations were applied to them. But these computational operations were different depending on what the relevant attitude was to the structure (and hence proposition). A belief structure plays a different computational role in the system than a desire structure, and these structures interact in ways specific for beliefs and desires, as prescribed in the schemas of the propositional attitude psychology. Fodor makes triumphant claims on behalf of his computational theory. He regards it as having achieved what Descartes claimed to be impossible: the mechanization of rationality (Fodor, 1983). One aspect of rationality is to connect what one wants and what one believes in a logical way. Kim's reaching out for the umbrella is rational in this sense. If the computational theory is correct, such rationality is fully covered by science, without any air of mystery remaining.

Fodor developed his "computational theory of mind" in a specific direction, very much guided by the ideas of Noam Chomsky. Assuming, like Chomsky, that the stimulus for language learning was impoverished, Fodor argued a child learner needed to be equipped with rather sophisticated reasoning capacities to figure out—by inferential reasoning—the structure of a language. He posited an innate "language of thought," or LoT (Fodor, 1975).

Functionalism, as a general approach in the philosophy of mind, needn't be committed to this language of thought. Neither need functionalism endorse the view of folk psychology as "propositional attitude psychology," or even the view of it as a theory. A functionalist can leave the task of defining mental terms and the causal relations between mental states to science. Such functionalist could leave open whether such a scientific approach to the mental would be in line with or diverge from folk psychology. They could even hold that there is no such thing as folk psychology.

Functionalisms aimed at scientific theories of the mind rather than folk psychology are called "psychofunctionalisms." The relevant scientific theories could be psychology, cognitive science, or computer science, or some discipline more like biophysics. All that a functionalist needs to be committed to are the following claims:

- the mental should be analyzed in terms of inputs, internal states, and outputs;
- it is the causal relations between inputs, outputs, and mental states that make a mental state into the state it is;
- because the same functional structure can be found in multiple material forms, the mind is in principle multirealizable.

8.4 The Foundational Myth of Functionalism

Functionalism became very influential in philosophy of mind and beyond. Functionalism's terminology and functionalist views have formed the framework for much philosophizing since the 1960s. Functionalism is accompanied by a specific interpretation of the field of philosophy of mind and of its history. There is a particular functionalist way of construing the philosophical positions that preceded it historically and of relating them in time.

This functionalist historical narrative merits attention. It provides an example of how thinking about philosophical problems and approaches can be shaped by the lens through which one is viewing. Functionalist construals of the positions before functionalism often do not provide an adequate characterization of what previous theorists intended to say and which problems they tried to solve. Rather, the established functionalist story tends to see previous philosophies as dealing with the same problems as functionalism, but less adequately. Functionalism then becomes the crowning achievement of a long-term development. A victor's story is told and it becomes widely adopted (Kuhn, 1962).

In his famous book on scientific change, *The Structure of Scientific Revolutions*, Thomas Kuhn notices how, after a fundamental change in scientific thinking in a discipline, the history of the science gets rewritten by the proponents of the newly established view (Kuhn, 1962). Previous episodes become seen as partial but still incomplete steps toward the novel approach. Kuhn also stresses the role of textbooks as means to consolidate and promote such history. Such a process happened with the cognitive revolution and the emergence of functionalism as a dominant force in the philosophy of mind. With some publications, such as Armstrong (1968), Putnam (1960), and Fodor (1975) in a leading role, a narrative about the development of philosophical thinking about the mind got created and repeated over and over—often partially with the very same phrases.

The general lesson that is illustrated by the functionalist retelling of history is that writing and thinking about philosophy doesn't occur in a vacuum. Philosophizing is an intellectual activity that is embedded in a particular time and place and that bears the marks of it. Disregarding these circumstances, or replacing them with a specific one, that maybe has just come into existence, can affect how one views older or other problems and positions. This is not something that can be avoided, and it is not necessarily bad. But there is a point at which a perspective becomes distortive. In the previous chapters, the positions that appear in the standard functionalist narrative have been presented in a way that tried to remain outside the functionalist framework. That will allow us to point at some of the inadequacies of that narrative.

8.4.1 The Narrative

It all begins with dualism, often taken to be a "natural attitude" toward minds, with Descartes at its pinnacle. Dualism is right about the special nature of minds, but it mislocates the source for it in a separate substance. Substance dualism is incompatible with the scientific worldview, in which only natural or physical processes can be invoked to explain what happens.

Then came behaviorism. To its credit, it is monistic, talking only about behavior or movement. But behaviorism is mistaken in reducing mind to movement. It only talks about inputs and outputs and thereby misses the inner aspects of mentality. That is, the mind as present only in subjective and private experience is completely overlooked by behaviorism. Moreover, cognition is not a reflex-like reaction to stimuli but involves stored knowledge or information.

The identity theory formed an improvement on dualism, because it is monistic. At the same time, it is superior to behaviorism, because it invokes the nervous system, thereby reintroducing the inner. But the identity theorists made a capital error by disregarding that the mind is multiply realizable. If a biologically different creature, or some being from a distant planet with a different physical makeup—a silicon brain, for example—could have a mind by having a similar functional organization, it is wrong to identify mental states with brain states.

Functionalism then combines all the advantages of the previous positions, by identifying mental states with functional states. It is monistic, it respects the inner, but in a nonchauvinistic way that allows for multiply realizable mentality.

8.4.2 Why the Narrative Is Mythical

For many reasons, the way functionalism conceives of history borders on fiction. It has already been questioned whether the behaviorism it invokes is merely a strawman. None of the philosophers usually named logical or analytical behaviorists have defended what is defined as logical behaviorism in the standard functionalist-leaning treatment. It is true that Wittgenstein, Anscombe, and Ryle opposed certain views of "the inner," but they fought certain conceptions of it, not the possible or actual existence of hidden feelings or unspoken desires. It has already been described how far their views were from aiming to reduce mentality to "mere movement."

Because the critiques of analytical or logical behaviorism are so widely reproduced, it is useful to make explicit why what goes for analytical behaviorism would not be endorsed by the philosophers that are, in the usual treatments, referred to as analytical behaviorists.

Two groups of philosophers who are regularly described as analytical behaviorists are, on the one hand, positivistic-minded people like Carnap and Hempel, and our "difference trio," Wittgenstein, Anscombe, and Ryle, on the other hand (see chapters 5 and 6).

Do they really qualify? Let's start with Carnap and Hempel. It is true that Hempel, on at least one occasion, described himself as a "logical behaviorist" (Hempel, 1935/1949). But, it is also true that both Carnap and Hempel's actual writings on the mind differ from "analytical behaviorism" in at least three respects. First, their treatment of mental terms included reference to the nervous system. The brain certainly counts as "inner" for the critics of analytical behaviorism. Second, their characterizations of mental terms and translations of psychological sentences were synthetic, not analytical. Carnap and Hempel shared this view with the identity theorists, as has been shown in detail in chapter 5. Third, Carnap made it very clear that the kind of characterizations and translations he gave were not meant to be complete. That is, as all characterizations, they hold in normal circumstances, pending unusual unforeseen conditions (see chapter 5).

What about Wittgenstein, Anscombe, and Ryle? Analytical behaviorism as construed in the functionalist narrative aims at giving reductive definitions for mental states that fully capture the meaning of mental terms. But such enterprise is possible only if one makes certain assumptions about language, for example, that words and expressions in a language have meanings that one can fully capture in definitions. This assumption might be valid in referential view of language. If words and expressions are essentially tools to refer to things, it is possible to provide an exhaustive characterization of meaning by pointing out what words refer to. Analytical behaviorism's core then consists of two claims: that mental terms refer and that they refer to movements or dispositions to make movements instead of to inner events. Both claims are incompatible with the writings of Wittgenstein, Anscombe, and Ryle. Indeed, their thinking starts from explicitly rejecting the first claim. Criticizing the view of words as essentially referential, name-like, and the picture-picture of language built on that assumption is at the heart of their philosophies. In rejecting the view that mental terms act as names, they renounce both that mental terms are names for inner items and that mental terms are names for movements or dispositions for making movements.

But don't they, in their nonreferential view of the mental, still disregard the inner? Consider that they want to bring the mental "back to life." They want to show that understanding puzzling mental concepts becomes possible when one considers the use of these terms in their surrounding circumstances. These circumstances involve other people, and other organisms, acting in certain ways. The contextualized activities of other persons or organisms allow for the proper use of mental language. One shouts out that one is in pain. Or one says that someone has lost confidence because one notices a hesitation in their eyes. The terms "pain" or "loss of confidence" do not refer to the activities of shouting or to the expression in the eyes. Rather the shouting and the look of the eyes form some of the circumstances that allow for the use of mental terms. Persons or organisms, their lives and histories, are part and parcel of these circumstances. Given certain, very complex yet specific circumstances, one can say that someone is suspicious, and a tiny change in the circumstances might lead one to qualify that assessment. Mental terms thus are tightly connected to aspects of the lives of persons. But aspects exist only with what they are an aspect of—they cannot be divorced from it like a smile cannot be removed from a face. Such thinking can be applied to "the inner." "Inner" or "private" aspects of mental life are aspects of personal or organismic lives too. They take place in the midst of, and can only be understood in the context of, the further circumstances of those lives. We can understand that someone, or some animal, suppresses pain, by finding a place for such suppression in their life. An animal doesn't want to show weakness to a predator; a Spartan child has been raised to show endurance and courage. Wittgenstein, Anscombe, and Ryle do indeed argue against a certain concept of the inner: the "inner" that is, or could be, completely separated from the lives and circumstances of persons and organisms. They hold that one can no longer make sense of mental life once it has been divorced in principle from the lives of persons and organisms.

Although, upon closer inspection, it becomes questionable whether logical behaviorism was actually held by the usual suspects, the rebuttal of analytical behaviorism was taken to imply that the concerns raised by Wittgenstein, Anscombe, and Ryle to long-standing philosophical views and methods could be left as they were. Philosophy could reconnect with many, or some, of the views that had been assumed valid by many philosophers throughout the development of Western philosophy. Such ideas include a referential or pictorial view of language, an object-like view of sensations or sensation-like entities, or the possibility of a unifying philosophical account that applies to everything from pains to emotions to beliefs.

What about the identity theory? The main issue functionalism has with the identity theory concerns multiple realization. The identity theorists are portrayed as remaining glaringly blind to the multiple realizability of the mental. In short, the identity theory is unduly chauvinistic.

In fact, the identity theorists were not concerned with multiple realizability. In the functionalist reconstruction, the identity theory is sometimes presented as making identifications on a structural basis, instead of on a causal-functional basis (Fodor, 1975, chapter 2). The identity theorists, for example, identified pain with processes going on in specific neural structures, as actually found in human brains, as when followers of Place and Smart talked about identifying pain with C-fiber firings (about the inaccuracy of this identification, see Pucetti, 1977). Functionalists then improve on this by making identifications on the basis of a causal profile.

But the structural/functional contrast is nowhere to be seen in the writings of the identity theorists. They write about brain processes, and that is open to interpretation in a functional/causal reading of "brain," according to which the notion of a silicone brain is coherent insofar as it shares its causal organization with biological brains. In fact, much of the identity theory was pitched exactly at the level of causes and effects. The main argument for making identifications between sensations and brain processes was that the brain process explained whatever the sensations explained. Such explanatory equivalence was unpacked in terms of causal processes: A brain process was taken to be identical to the greenish afterimage because the brain process was similar to brain processes caused by green things, and because the brain process caused the same introspective judgments that were caused by the afterimage experience.

The relation of functionalism to dualism isn't straightforward either. Like all their predecessors, functionalists tend to reject dualism. Yet in some ways, functionalism makes the opposite movement. Several functionalists defend functionalism only insofar as it concerns the cognitive aspects of mentality. David Chalmers, for example, defends a functionalist and computationalist view of cognition, while at the same time arguing that functionalism cannot deal with the phenomenal aspects of consciousness (Chalmers, 1996). These arguments will be treated in the next chapter. In the meantime, it should be noticed that, also on the topic of cognition, or of "cognitive mental states," functionalism retains strands of thinking that have always formed the backbone of dualism. As detailed in chapter 2, historically, dualism was motivated by the idea that the abstract nature of thought made it incompatible with matter.

Consider multiple realizability again. Functionalists assert that the mind is multiply realizable. In principle, alien creatures could think or feel pain, even if they would have nonbiological bodies, so Putnam claimed when he launched the idea of multiple realizability in the philosophy of mind. But why would one believe that multiple realizability holds in general for cognition? The claim about the multiple realizability of the mental is an a priori claim. It is not made based on the reliable observations of many concrete cases of multiply realized nonorganic mentality. Rather, it is motivated, on the one hand, by intuitions that aliens could have pain, or could think, and on the other hand, by highly theoretical considerations about computation and information processing (see, e.g., Clark, 2014, p. 25).

Some things are indeed multiply realizable: One can make a pen in plastic, wood, or metal. Some parts of our body can be replaced by artifacts. It is common practice to replace the eye's natural lens after some cataract operations with an artificial intraocular lens. It can be made of acrylic, or silicone, or other plasticlike materials. But if the material changes, some properties might change too. Your metal pen will be lost when you drop it in a lake, but you might recover your wooden pen. So, whether something can be realized in a different material depends on which aspects you want preserved in the new realization. A different realization will show some aspects of the original, but not other ones. Further, whether something can be multiply realized depends on what it is. Perhaps pens can be easily realized in many kinds of materials, but what about strawberries, or the Martin's 0x2e Cocobolo-Acoustic-Electric Guitar? The way these are made, from which specific materials, seems to be unique. If one changes the material, one will have different properties, among which a different taste and a different sound. That is, why would one assume, just based on the claim that it's easy to imagine Martians could have pain or hold beliefs, that the mind or cognition is multiply realizable?

Here, ideas about information processing and computation can help to motivate the actual multiple realizability of the mind, or at least of cognition. Information seems to be independent of what carries the information. You can signal your presence on my doorstep by a knock, a ring, or perhaps a howl. If information can be present in many substrates, so can its processing. And the same reasoning can be applied to computation. Or one could ground the multiple realizability of the mind in the thought that the components of brains, that is, neurons, are a sort of calculating device. Neurons are activated by other neurons, and as a consequence become more or less active themselves, influencing other neurons in turn. If these activations are quantified, it's all a play of numbers coming in and going out, and the same numeric patterns could be implemented in a device made of silicon neurons. Neurons are calculators, hence, they are computers.

These ways of motivating multiple realizability come down to considering the brain or the cognitive system as engaging in something abstract: information processing or computation. It is because its underlying processes are abstract that cognition or mentality can be multiply realized. Continuous with dualist traditions, from antiquity to Descartes, the abstract gets highlighted as the key feature of the mental. As in the older dualist traditions, in functionalist philosophy, the abstract nature of thinking is opposed to the concrete nature of what thought is embodied in: the biological brain and body.

It is important not to lose out of sight that an abstract characterization is just that: an abstract characterization. A computational characterization of neural activity is a model of what neurons do rendered in a descriptive system devised by humans, such as that of numbers and arithmetic. To say that the neuron implements the model invites to turn things on their head, to assume the model was there first, for the neuron to later implement. Of course, when humans build a calculating machine, they first have a design that models what the machine should do. When the calculating machine is constructed, its functioning is compared to its design. But this is not how nature and evolution work, at least not on a naturalistic understanding. There are no models in nature that precede how things function or evolve. There is no blueprint for reality.

In reality, concrete things are primary. Bodies, brains, and neurons exist and function, but not based on preexisting blueprints. Instead, models for bodies, brains, and neurons come after the particular bodies, brains, and neurons. Any model, for a neuron or whatever, is always incomplete and idealizing with respect to what it models. Models do not consider all aspects, and they approximate.

There is an important distinction between the claim that a model for mentality can be multiply realized and the claim that mentality can be multiply realized. The first claim could be true while the second isn't. A computer can run an implementation of a model for the weather, but it thereby does not implement the weather. Similarly, a computer might run an implementation of a model for some cognitive or mental process, but it would not thereby implement the cognitive or mental process. To equate the multiple realization of models of mind with the multiple realization of mind is the confuse the model with the modeled.

8.5 Eliminativism and Intentional Systems

Though the functionalist view came to be widely held in philosophy, it did not remain without its challengers. One noted strand of criticism drew from a revival of eliminativism—the position developed in the line of the early eliminativisms of Paul Feyerabend and Richard Rorty (see chapter 5). The main defenders of the new eliminativism were Paul and Patricia Churchland (see Churchland, 1981, 1986).

The Churchlands' eliminativism agreed with Sellars, Lewis, and many functionalists that folk psychology was a theory. But contrary to leading functionalists such as Fodor, the Churchlands argued that it was a bad theory (Churchland, 1981, offers a pointed presentation and defense of this claim). In a move explicitly reminiscent of Feyerabend, they claimed folk psychology—thought of as a kind of functionalist theory—was no better off than other disreputed theories, such as the witch theory of diseases, the caloric theory of temperature—according to which temperature was a kind of stuff, like a fluid—or the ether theory of light—according to which light needed a medium to travel in. These theories are now replaced by the biology of bacteria and viruses, the thermodynamics of molecular motion and the theory of relativity, and it is no longer believed that the entities they used to explain phenomena with exist. There are no witches, no caloric, and no ether.

The Churchlands' eliminativism targeted propositional attitudes in particular. Again agreeing with Fodor, they saw folk psychology as built on the assumption that beliefs, intentions, and hopes were internal states posited to explain behavior. But while Fodor praised folk psychology for having discerned correctly what causes behavior, the Churchlands made the opposite assessment. Alleged explanations in terms of propositional attitudes, the Churchlands argue, are superficial and wrong. Just like there's nothing real that corresponds to caloric, there is nothing real that corresponds to "The belief that it will be warm tomorrow." Just as caloric talk and theory were replaced by thermodynamical talk and theory, so too will propositional attitude talk and theory be replaced. The Churchlands point to neuroscience as what will do the work thermodynamics did with respect to caloric. Based on connectionism and developments in neuroscience (Churchland, 1986, 1992, 1995), they claim neuroscience is positing numerical characterizations of brain processes, instead of propositional ones. Brain science, in the Churchlands' analyses, talks about levels of neural activation, and neuroscience's regularities concern how these activation patterns evolve within and across groups of neurons. In using a numerical rather than a propositional vocabulary, neuroscience conforms to other natural sciences, such as physics, chemistry, or biology. In none of these sciences does one find equations relating propositional attitudes, and neither should one hope for such in the mature neuroscience that will come to replace folk psychology.

Despite offering a profound critique of assumptions widely held by contemporary colleague philosophers allied to the cognitive science agenda, the Churchlands' call for reformation did not touch upon the most fundamental pillars of the cognitivist framework. They held on to the tenets of representationalism, computationalism, and internalism. They agreed with other cognitive science–inspired philosophers that cognition was all about building and transforming representations by means of computational processes. They held on to internalism because they held these processes were located in the brain. Their challenge concerned the format of these representations. In their views, the format of brain representation was numerical. Brains represent not by means of the sentence-like parts of propositional attitudes, but by means of numerical neural patterns. Such neural patterns could be distributed patterns of activation of groups of neurons, or the strength of the connection weights between neurons. Indeed, Paul Churchland very explicitly portrays connectionist concepts as plausible replacements for folk psychological terms (Churchland, 1989, 1995).

Daniel Dennett developed a position concerning folk psychology and the propositional attitudes that diverges from both Fodor's functionalism and the Churchlands' eliminativism. Dennett held that Fodor was excessively realist with respect to beliefs, desires, and the other propositional attitudes—he coined the term "industrial strength realism" for Fodor's position (Dennett, 1991b). The Churchlands, on the other hand, went way too far in aiming to eliminate all talk of beliefs and desires. Both parties, according to Dennett, wrongly assessed what using propositional attitude talk commits one to. They both assume that propositional attitude talk can be vindicated only if propositional attitudes exist as real states inside organisms. Fodor then claims they do exist, and the Churchlands deny that. So, contra Fodor, Dennett held that one cannot expect that propositional attitudes map one-to-one into brain states. Yet, contra the Churchlands, Dennett also held that propositional attitude talk is extremely useful and successful. Dennett accepted Fodor's claims that folk psychological explanations proceed by ascribing beliefs, desires, and other propositional attitudes to explain and predict behavior, and that this strategy is very effective. Yet Dennett argues one should not commit to Fodor's ultrarealism about them. However, the fact that there will be no propositional attitudes found in the brain, should not lead to eliminativism.

According to Dennett we should understand propositional attitudes in terms of the possibility of taking "the intentional stance" (Dennett, 1971,

1987). One takes the intentional stance when one considers a system as something that has goals it tries to reach, and beliefs it relies on in trying to reach these goals—and something similar is meant to apply for other propositional attitudes. In principle one can take the intentional stance towards anything. One can, as to some extent did Aristotle, ascribe the goal of being nearer to sea level, to a stone rolling downhill. But, as the history of science has shown us, it is wiser to treat the stone as a physical system—one only subject to physical laws. Yet for evolved systems like organisms, taking the intentional stance pays dividends. Thus, to some extent, any evolved organism will have the goals it has to have and believe what it has to believe. The intentional stance makes those beliefs, desires, and, by extension, other propositional attitudes visible and allows to rely on them to explain and predict the organism's actions. Suppose the feeding bowl of a dog has been moved to a new location, unknown to the dog. One sees the dog on its way to the old location. It would be far-fetched to explain its behavior in physical terms, or by taking "the physical stance." Even if such would be possible in principle, this would require massive knowledge about the physical structure of the dog's brain, body, and interaction with the environment. Taking the intentional stance allows to explain its behavior easily and accurately: The dog's walking is due to the combination of its desire to eat and the belief that the bowl is at the regular location.

Evolution offers the reason the intentional stance works. It is because of evolution that organisms have propositional attitudes, and it is because of evolution that propositional attitudes are related in certain rational and predictable ways. As evolved living beings, organisms must have goals which they successfully achieve in knowledgeable ways. If they didn't attempt to stay alive to keep away from predators, and if they didn't register and put to use relevant information in their surroundings, they wouldn't survive.

So, contra the Churchlands, and with Fodor, Dennett holds that ascribing beliefs, desires, and other propositional attitudes allows for the accurate explanation and prediction of behavior. But, contra Fodor, he insists that one should not jump to the conclusion that, therefore, beliefs and desires are states of the brain, or of the cognitive system. That is, according to Dennett, whether the success of propositional attitude talk can be explained only by resorting to the excessive realism of Fodor, as Fodor himself claims, is something that is up in the air. Dennett is cautious about Fodor's claim that the generative nature of cognition can *only* be explained via assuming the existence of a language of thought. Dennett points out that in fact no other system with the generative properties of human cognition is known but human language. But he doesn't take this as a sufficient reason for endorsing the Fodorian claim that cognition is underwritten by a language of thought. Dennett writes:

> So the argument for a language of thought comes down to this: what else could it be? We have so far been unable to imagine any plausible alternative in any detail. That is a good enough reason, I think, for recommending as a matter of scientific tactics that we pursue the hypothesis in its various forms as far as we can. But we will engage in that exploration more circumspectly, and fruitfully, if we bear in mind that its inevitable rightness is far from assured. One does not well understand even a true empirical hypothesis so long as one is under the misapprehension that it is necessarily true. (Dennett, 1987, p. 35)

Though Dennett's approach to the language of thought is more cautious than Fodor's, he still subscribes to the idea that it is probably true that *some* generative system underlies human cognition (Dennett, 1987).

What exactly is Dennett's position? In section 3 of this chapter, two ways of understanding propositional attitudes were discerned. According to one, ascribing a propositional attitude comes with an assumption that the subjects themself hold in mind some kind of description corresponding to the proposition in the propositional attitude. In the second construal of propositional attitudes, this assumption is not made. Dennett's intentional stance seems to reject the first interpretation. But then his commitment to some unspecified generative system behind cognition seems to open a backdoor for some version of this interpretation to reenter.

Consider this example of a propositional attitude ascription: We say of the dog that it believes that its boss has entered the house. Given Dennett's critique of the Fodor, one would think Dennett holds that the ascription of this propositional attitude to the dog is not best understood as involving the dog standing in a relation of belief to a token of a sentence type in its language of thought. But how does this relate to Dennett's assumption that there is some generative system behind the dog's behavior? Does this system, or does it not, allow the dog to picture or describe in some sense that its boss has entered? What is this system? How is it generative without being a language? We have an idea what generativity means in language: You can form sentences at will by combining words. But what does generativity mean in that other, unspecified system? Intentional system theory remains silent on questions like these. Because intentional systems theory remains underspecified and schematic, it is hard to assess what answers it provides to specific questions.

Dennett also retains another assumption made by Fodor, by many functionalists, and by the Churchlands. Dennett agrees that there is something like a "folk psychology" and that its main means to explain and predict behavior are the propositional attitudes. According to this story, when we explain why we see the dog moving to its food bowl, or Uncle Pete to the fridge, we invoke beliefs and desires: a desire to drink, and a belief there is water in the bowl and juice in the fridge. One can question whether such a construal of dealing with each other's actions isn't overly schematic and very partial. For one thing, in our everyday dealings with each other, and the uses of mental vocabulary, are our main aims "theoretical," like the aims of a scientist? Are we really focused on prediction and explanation, rather than on influencing other people, steering expectations, and regulating our social relations with others? As per Anscombe, when one expresses an intention, one expresses one's commitment to a course of action in a practical context. This might allow the other person to fine-tune their course of action—preparing for an extra guest at the party, for example.

Bibliographical Note for Chapter 8

David Armstrong provides some reflections on his broad use of the term "state" in section 1 of chapter 7 of *A Materialist Theory of the Mind* (Armstrong, 1968).

Contemporary computationalist philosophy has added many twists and nuances to the earlier version. For example, computationalism proposed and defended in Piccinini (2015) is built on a syntactical notion of computation—one defined without having recourse to semantic properties. Nevertheless, contemporary computationalists defend a representational view of cognition.

Though philosophy has widely embraced "multiple realizability" from early on, some dissenting opinions and criticism can be found in Polger and Shapiro (2016) and, very forcefully, in Zahnoun (2023).

9 Feeling and Meaning

> I can easily distinguish between a notion and a sensation. It is one thing to say that I have the sensation of pain. It is another thing to say that I have a notion of pain. The last expression signifies no more than that I understand what is meant by the word *pain*. The first signifies that I really feel pain. . . . Every man agrees . . . that to suffer pain, and to think of it, are things which totally differ in kind.
>
> —Thomas Reid (1785/2002, pp.158–159)

9.1 From Phenomena to Problems

It has become common to structure the philosophy of mind around two central problems. One of these is the problem of phenomenal consciousness. The second is the problem of intentionality. This overall view of philosophy of mind goes hand in hand with the idea that there are two properties that characterize the mind and at least seem to set it apart from everything else in reality (Bayne, 2022; Kim, 2011). The first is phenomenal consciousness: that experiences feel a certain way. Many philosophers have argued that phenomenal consciousness has purely qualitative aspects—for example, how pain feels—that cannot be understood in the same ways that we make scientific sense of the natural world. These purely qualitative aspects, so some philosophers argue, cannot be explained if one holds fast to philosophical views of the mind such as the identity theory or functionalism, or any other form of thinking labeled as "physicalism." Some have argued that the purely qualitative properties of consciousness require revisions in our overall worldview, revolutionary scientific progress, or both.

Next to phenomenal consciousness stands intentionality. Intentionality in this context should be distinguished from the intentionality associated with things that are done intentionally, an issue central to Anscombe's *Intention* (see chapter 6). Intentionality in the sense we are concerned with in the current chapter means that something is directed upon something else. Perception, imagination, thought, and, according to many, everything mental reaches out to something else: perception to what you perceive, imagination to what you imagine, and thought to what you think. But how can something that occurs at some place and time reach out to something else at another time and place, or perhaps something without a time or place, as when you imagine the nymph Circe from Greek mythology? Some philosophers have taken the oddity of intentionality to pose a problem for naturalism, just in the way phenomenal qualities might do.

It is important not to take for granted that phenomenal consciousness and intentionality form timeless and particularly tough philosophical challenges. For example, both Aristotle and Descartes wrote about conscious experience, but neither of them argued that the special nature of sensory consciousness provided reason to question that we could understand experience as natural phenomena (see chapter 2). Both Aristotle and Descartes did question whether minds could be understood like other natural phenomena, but the basis for this assessment lay in the rationality of minds, not in the existence of experiential qualities.

Moreover, the turn toward phenomenal qualities and intentionality and many of the philosophical arguments resulting from it would have been highly suspect for a lot of the thinkers we have dealt with in previous chapters. Neither Reid, Carnap, Place, nor Anscombe would have endorsed the assumptions fueling the now familiar "two outstanding problems" view of philosophy of mind. The turn toward phenomenality and intentionality is comparable to the developments that constituted the cognitive revolution (see chapters 7 and 8). In both cases, theorists act as if they regained freedom to investigate ideas that were previously suppressed. In the case of cognitivism and its associated philosophy, internal cognitive processes have been reconquered. In the case of the philosophy of consciousness, the special conscious properties of experiential states are liberated for philosophical exploration.

Let's see in some more detail how this situation came to be, beginning with phenomenal, or qualitative, consciousness.

9.2 Feeling and Qualia

What is the relation between the world as it appears in perceptual experience and the world as studied by the natural sciences? Very different answers have been given in the development of Western philosophy. According to the Aristotelian view, that was dominant, even if not uncontested, from antiquity until the scientific revolution of the sixteenth and seventeenth centuries, perceiving properties of objects came down to sharing forms with it. When a person perceives a lemon as yellow, the person comes to share the form of yellow that was in the object. Yellow now informs both the lemon and the perceiver. There were dissenting voices in antiquity. "By convention sweet and by convention bitter, by convention hot, by convention cold, by convention color; but in reality atoms and void," Democritus wrote, implicating that sensory properties such as colors, sounds, and tastes were in the perceiver rather than in the objects perceived (Democritus, 1951/1999, DK68B9). But such doctrines as espoused by Democritus were kept outside of mainstream thinking, that remained Aristotelian. During the sixteenth and seventeenth centuries, the Aristotelian worldview was replaced by the mechanistic picture. Natural processes no longer were explained as flowing forth from inherent forms, but as resulting from how different material bodies moved and influenced each other's movement.

The view that at least some sensory properties such as color were effects in the mind that were unlike their causes became widely accepted. One can see it at play in Descartes (see chapter 2) and in the notion of secondary qualities and in psychophysics (see chapter 3). In these conceptions, there is a gap between the properties of the world as studied by physics and how these properties show up in conscious experience. What are wavelengths for physics are colors in the mind.

However, sensory qualities were not a basis for Descartes's defense of dualism. His dualism was founded on rationality. In Descartes's view, sensory properties become conscious only if accompanied by thought. Thought is the basis for the consciousness of sensory properties, not the other way round (see chapter 2, section 2.3). By the second half of the twentieth century, the role of sensory properties in views on the relation of the conscious mind and the physical world had changed.

The special nature of sensory properties became a self-standing basis for arguing against broadly naturalist views of the mind—views according to which the mind is a natural phenomenon that can be understood like other natural phenomena. Experiences of color could be invoked to argue against naturalist views of the mind, without mentioning rationality. So qualitative phenomenal consciousness took the place of rationality in arguments against the possibility of understanding the mind in the ways science understands the natural world of quarks, stones, flowers, or bees. That rationality ceded its place will have been helped by the growing confidence that the cognitive sciences were making great progress in mechanizing rational thought (see chapter 7).

In particular, it became common to problematize physicalism about the mind by referring to the differences between experiential properties and physical properties. According to this line of reasoning, experience is thought to have purely qualitative properties, which seem to have nothing to do with physical causation. When you experience a color or are in pain, there is a purely qualitative aspect to your experience that cannot be elucidated by how the experience is caused or what effects it has. Physiological, neurobiological, or functional knowledge of your experience can be acquired, but it will never touch upon this purely qualitative aspect. Therefore, because experience comes with purely qualitative aspects, views of the mind as physical are in trouble.

An extremely influential articulation of this "qualitative feel" argument against physicalism was given in Thomas Nagel's article "What Is It Like to Be a Bat?" (Nagel, 1974).

In assessing the fortunes of physicalist views of the mind, which included behaviorism, the identity theory, and functionalism, Nagel considers the perceptual experience of a bat. Bats echolocate. They sense distance by emitting a sound (a shrill), which bounces off surfaces in their environment and returns to the bat. Because of the difference between the time of emission and reception of the sound, the bat can become aware of the location of things in its environment. Echolocation is neither hearing nor seeing. Throughout his paper, Nagel assumes the bat's perception by echolocation is experiential. But then the question is: How does echolocation feel to the bat, or, in the influential turn of phrase used by Nagel, "What is it like to be a bat?" (Farrell [1950] used the same phrase). In Nagel's analysis, the trouble for physicalist views is that no matter how much scientific knowledge one has, or ever will have, about the bat's brain and its functioning, the "what it is like" question remains unanswered. Physical knowledge does not seem to bring "what it is like" knowledge.

In "What Is It Like to Be a Bat?" Nagel does not go on to draw more than a conclusion about understanding. He does not say the mental is not physical, only that physical knowledge falls short in allowing us to understand the phenomenal aspects of experience. Though Nagel's paper has been extremely influential and has been taken to provide crucial insights for the philosophy of consciousness, there is a straightforward way, at least as old as the writings of Thomas Reid, of responding to it which blocks its grave philosophical consequences. This is to point to the difference between "having an experience" and "knowing about an experience" (see the epigraph for this chapter). Obviously, a bat has experiences and then feels like a bat feels. But to say that, by feeling its feelings, a bat also *knows* what it is like to feel requires an extra step. It is the step of assimilating experience and knowledge about experience. But perhaps a bat just experiences without in any sense knowing what it is like to experience its experience. Perhaps, thanks to our access to science of how bats perceive and act, we people are more entitled to having knowledge of what it is like to be a bat than a bat has (Akins, 1993; Dennett, 1991a, pp. 441–448). Of course, one will never experience what a bat experiences by becoming knowledgeable about bats. There is a notion of being a bat that one can have when one is not a bat, but this should be distinguished from *being* a bat—just as there is a difference, to give Reid's example of the epigraph, between thinking about pain and being in pain. I can think as hard and detailed as I can about chocolate, but that will always be different from actually having a piece of Belgian fondant in my mouth. In Reid's terms, this seems to be very much something to which "every man agrees."

Put in another way, one can accuse something of falling short only if there is a reasonable expectation that it could not have fallen short. The most minimal condition for blaming someone for not achieving a result is that the result could possibly be achieved. If Nagel wants to show that physicalism is incomplete in certain respects, it should have been possible for physicalism to be complete in these respects. Nagel admits to having no clue about what such completeness would in practice amount to, or what would be or come near to an example of the missing description. Yet, we *can* say in the abstract what achieving such completeness would involve. It would involve leveling the difference between knowing what it is like to be a bat and being a bat, or between the notion of pain and feeling pain, between a description of an experience and the experience itself.

There is one other paper in contemporary philosophy of phenomenal consciousness that can lay claim to equal prestige as Nagel's (1974). This is

Frank Jackson's paper "Epiphenomenal Qualia" (Jackson, 1982). By the time Jackson wrote the paper, it had become common to refer to the "what it is like" properties of experience as "qualia." So, when bats have echolocation experiences, they have certain specific qualia, just like we have pain qualia when we feel pain. Interestingly, Jackson's paper can be seen as attempting to achieve exactly what was diagnosed as the central task that Nagel would have had to accomplish to show the limitations of physicalism or naturalism. Frank Jackson's leveling operation pivoted around the notion of "facts." It was the notion of "facts" that allowed Jackson to rhetorically bridge the gap between, and eventually equate, having a notion of an experience and having the experience.

Jackson's master argument is known under the name "the knowledge argument." Nagel argued only that we cannot physically understand the phenomenal without drawing the further consequence that therefore the phenomenal is not physical. Jackson, however, takes that further step. He argues that, because physical knowledge does not deliver knowledge about the purely qualitative aspects of phenomenal experience, or qualia, phenomenal experience or qualia cannot be physical (Jackson, 1982). How does the argument run? Jackson imagines a super scientist, Mary, who knows all there is to know scientifically about the physical aspects of color and color perception. All the optical and (neuro-)physiological facts about color are known to Mary. Yet, we are asked to assume that Mary is confined to an environment in which she only sees shades of gray. So, she has never seen a color until, one fine day, she leaves her gray world and sees a red apple. At that moment, Jackson claims, Mary comes to know a new fact: how it feels to see red. Because she previously had all the physical knowledge and upon experiencing red comes to know something new, Jackson infers, there are facts that physical knowledge provides no knowledge of. Therefore, some facts are not physical. Or so goes Jackson's much-discussed "knowledge argument."

A response to Jackson inspired by what Reid writes would be to point out that Jackson is misleadingly labeling experiencing as a form of acquiring factual knowledge. On this view, the whole argument relies on assuming that "coming to known facts" applies in the same way to reading books about neuroscience and to having an experience. But this erases a difference on which "every man agrees," namely the difference exemplified in thinking about pain versus suffering pain.

The point can be made by supposing, for the sake of argument, that a general form of the mind/brain identity theory is true—applying to sensations and other mental phenomena (see chapter 5). Suppose that for Mary to experience red, a specific brain process, R, must occur. For Mary to acquire the knowledge that experiencing red corresponds to brain process R, another process, KR, must take place. Under these conditions, Mary's experiences would be entirely physical. Yet, despite knowing all the physical facts while inside in the room, she would still encounter something new upon seeing color for the first time outside the room. Crucially, experiencing something new is here not described as "coming to know a new fact."

Peter Godfrey-Smith is a contemporary philosopher who has criticized the distinction Jackson's argument relies on in ways reminiscent of Reid's observation. He complains about views of the qualitative aspects of experience such as Jackson's in the following way:

> Some demands made of the materialist become unreasonable. A scientific description can't encapsulate or *contain* an experience being described; knowing about an experience is different from *having* it, even if knowing about it might help you imagine it. Some critiques of materialism seem to want a third-person description of a human or other animal to do something it could not, in principle, do: they want it to magically become first person.
>
> When people worry about qualia, they sometimes also set things up like this: the materialist is supposed to describe a collection of physical processes from the third person, and do so in a way that conjures up qualia. The reds and greens and sounds of cymbals are supposed to somehow *appear* in the system being described. This is entirely a mistake. (Godfrey-Smith, 2020, pp. 109)

Godfrey-Smith goes on to make a point in line with how Place and Smart motivated the mind/brain identity theory: To explain why an organism that answers a certain description enjoys conscious experience, one only needs to invoke that the organism *is* the one to which the description applies. Being that organism suffices; there is no further mysterious fact or property that is left out of the equation. As Godfrey-Smith writes:

> The qualia are not extra things that need an explanation, somehow produced by the workings of the physical system. Instead they are part of what it is to *be* the system described. Experience is the first-person point of view of a complex living system of a certain kind, not something conjured up by the workings of that system. (Godfrey-Smith, 2020, pp. 109)

Apart from the fact that the identity theorists more narrowly talked about "having a brain process" where Godfrey-Smith writes in terms of "being a system," the same philosophical claim is made by both parties.

Godfrey-Smith's line of reasoning also connects to Ullin Place's invocation of the example of a cloud and the collection of droplets out of which it is composed (Place, 1956, see chapter 5). Place observed that we can be related to a cloud in different ways, depending on whether we are inside or outside of it. Place warns about mistaking these two ways of relating to the cloud for one way of relating to two different things. Adapted to the current context, this would lead to saying that Mary can be related in different ways to the same facts. She can be related to the physical conditions under which she experiences color by reading about and intellectually coming to know these conditions, and by having these conditions applicable to her—by being in the conditions described by what she's reading; by what the knowledge is about, being applicable to her. These different ways of relating to the same conditions are mistaken by Jackson as one way of relating to (knowing) two different conditions (physical and nonphysical).

David Lewis has pointed out that the knowledge argument can be addressed to dualism as well as to physicalism. He wrote:

> Let parapsychology be the science of the non-physical things, properties, causal processes, laws of nature and so forth that may be required to explain the things we do . . . Black-and-white Mary may study all the parapsychology as well as all the psychophysics of colour vision, but she still won't know what it's like. (Lewis, 1988/1999, p. 289)

This corroborates that what the knowledge argument revolves around is not a distinction between fundamentally different categories of physical facts and experiential facts, but a difference between having notions and having experiences, or between different ways of relating to the same reality.

There is a further difficulty that besets anyone endorsing the knowledge argument. If it is concluded that qualia are nonphysical, they become noncausal. That is, they become what is known as "epiphenomena." An epiphenomenon is a phenomenon that is caused but doesn't cause anything itself. A famous example of an epiphenomenon, due to Thomas Huxley, is the sound of a steam whistle made by a locomotive. It is caused by the engine but plays no further role in the production of the train's motion (Huxley, 1874).

Jackson is happy to endorse the view that qualia are epiphenomenal (Jackson, 1982). They must be, according to the logic of the knowledge argument, because if they weren't, they would be fully part of the causal fabric of the world and so would be covered by the totality of scientific knowledge and black-and-white Mary would have knowledge about them.

If qualia are epiphenomenal, this means that they don't have effects themselves. This seems problematic because it seems that experiences do have effects. When you say the chocolate tastes great, or if you withdraw your hand because you feel pain upon touching a flame, you seem to react to the experience, and to react in a specific way because of the specific quality of the experience. You are delighted by the taste of the chocolate and distressed by the burn.

Jackson bites the bullet and argues that qualia are epiphenomenal nevertheless. The reason he offers is that it is not unthinkable that the qualia are epiphenomenal. It remains a possibility that the reactions to the chocolate or the pain are in fact reactions, not to the qualia, but to an underlying physical process that causes both the qualia and the reactions (Jackson, 1982).

Jackson's strategy is one of attaching great importance to conceivability. A prima facie theoretical option, namely, that qualia are epiphenomenal, is seen as a serious contender among other theoretical options because it is possible to coherently conceive that the option would be true. Conceivability became central to the further development of qualia-based thinking about consciousness that was skeptical about the fortunes of physicalism. David Chalmers probably is the most prominent representative of this line of thinking. In his influential book, *The Conscious Mind: In Search of a Fundamental Theory*, Chalmers argued that, because of qualia, physicalism faced a particularly difficult "hard problem of consciousness." The hard problem of consciousness is to account for the existence of qualia in a physical world by the theoretical means available in the standard sciences of that physical world (see also Levine, 1983 for a famous similar diagnosis in terms of the existence of an "explanatory gap").

By the time Chalmers (1996) appeared, the philosopher Ned Block had already resurrected Locke's inverted spectrum thought experiment to argue that functionalism could not account for color qualia. According to Block (1978), two persons could function the same, yet have the qualitative aspects of their color experiences inverted. Where the one experiences red, the other experiences green; likewise with blue and yellow, and every other pair of hues. Despite the qualitative differences, one can imagine that the two would function exactly alike. They both would call a ripe tomato "red" and say it looked like a ripe strawberry. Following Block, Chalmers argues that the conceivability of spectrum inversion brings functionalism into trouble because the inverts would be exactly the same when considered by functionalism.

Functionally, their mental states for "red" are the same. Yet phenomenally these experiences differ because they have different qualitative aspects. Therefore, the phenomenal slips through the maze of functionalism.

With similar reasoning, one can invoke philosophical zombies (Chalmers, 1996). Just like color inverts are conceivable, this line of reasoning goes, so are persons that function like persons with phenomenality yet without actually feeling anything. These are philosophical zombies. Again, because functionalism doesn't make any difference between a person with phenomenal feel and a zombie; it misses what makes the difference between the feeling person and the zombie: the phenomenal. The zombie argument can also be directed more broadly against any form of physicalism, not just functionalism. Philosophical zombies then have all the functional and physical properties of an original person but lack their qualia (see, e.g., Chalmers, 1996).

Chalmers (1996) and others have called for a nonphysicalist metaphysics, based on arguments involving spectrum inversion and philosophical zombies. One idea that Chalmers has explored is that of a "naturalistic dualism." It would introduce phenomenal properties alongside the currently accepted fundamental properties of physics. Qualitative consciousness would then become a fundamental property of nature that would be lawfully associated with certain physical or functional properties. For example, fundamental pain qualia would be correlated with neural pain processes.

Recently, some philosophers have explored a different nonphysicalist metaphysics, by reviving the idea of *panpsychism* (Goff, 2019). According to panpsychism, some form of qualitative consciousness is present throughout the universe. According to panpsychists, qualitative consciousness does not—indeed, *cannot*—emerge from something that completely lacks it. Therefore, if qualitative consciousness has emerged from physical matter, matter must be endowed universally with protophenomenality, or so the panpsychist line goes.

Panpsychism, Chalmers's "naturalistic dualism" or Frank Jackson's notion of epiphenomenal qualia all claim that the aspects of reality allegedly left out by the positions they criticize have a special metaphysical status. Neither of these authors provide detail of how nonphysical qualia operate. Chalmers claims that qualia do all their work via their correlating physical processes, and the same seems to be true for Jackson's epiphenomenal qualia. No account is given, by either of these authors, of how qualia came to be, why they have come into existence, or how one can explain the allegedly purely qualitative difference between the quale of seeing red and the quale of hearing the sound of a bell (O'Regan, 2011; see also Reid's response to Hume's example of seeing a table in section 3.4.4 of chapter 3). In this respect they might be compared to Descartes's writings about dualism (chapter 2). Descartes pointed at the other substance as the substratum of thinking but never gave much detail about how the thinking substance worked in its own special, nonmechanistic ways, while he proposed a multitude of theories regarding the supposed mechanistic processes underlying physical phenomena.

The reliance on conceivability also shows similarities to Descartes's rationalism. For example, Chalmers argues from the fact that it is possible to conceive of a philosophical zombie to the existence of a special kind of nonphysical properties. This requires strong confidence in one's reasoning powers. Wouldn't we need to investigate reality scientifically, empirically, to discover what it is made of? Isn't that how we found out about electrons, photons, and bosons? The idea that we could discover just by thinking what are reality's fundamental constituents is reminiscent of Descartes's claim that he could, by reasoning from clear and distinct ideas, acquire certain knowledge about the ultimate composition of reality.

9.3 Meaning and Content

Philosophers of mind have been worrying about more than phenomenal consciousness. Another aspect of the mind that has puzzled them for many decades is what is known as *intentionality*. "Intentionality" in the sense relevant here should be distinguished from the intentionality or purposefulness of actions that are done with an intention. That topic was discussed in chapter 6. Though it is possible to relate the purposefulness of action to the sense of intentionality discussed here, it is best to keep the notions separate.

Intentionality as understood here is a general property of "being about." When I say, standing in a room peeking out of a window, "It is raining outside," I utter my thought in the room, but what the thought is about is something outside. My thought "reaches out," and manages to be about something at a distance.

Philosophical discussions about intentionality go back many centuries. The word "intentionality" is supposed to be inspired by an analogy with the stretching of a bow—one of the meanings of the Latin "intendere" (Jacob, 2023). The concept also came up in the context of early scientific psychology and would become central to the philosophical movement of phenomenology, initiated by Edmund Husserl (1859–1938). Husserl in his turn was inspired by Franz Brentano (1838–1917). Brentano had declared intentionality to be a "mark of the mental": a property that is uniquely specific for the mental (Brentano, 1874). Brentano claimed that all consciousness is consciousness of something. Consciously perceiving, thinking, hoping, or imagining always has an object: that what is perceived, thought, hoped or imagined (Brentano, 1874). What is perceived can exist, but it needn't. Just as one can think about the rain outside, one can think about the mythical nymph Circe. But now one can ask: How is this reaching out possible? How can the thinking in the room travel through space to reach the rain outside, or even some nonexistent mythical creature? Brentano went so far as to hold that, because of its intentional nature, the mind could not be physical. It couldn't be, because intentionality allows the mind to transgress the boundaries of space, time, and possibility that constrain physical interactions. In this way, intentionality becomes similar to phenomenal consciousness. It becomes a property for which it is allegedly impossible to give a naturalistic account (see Bayne, 2022, chapters 1 and 6).

But one needn't construe intentionality as potentially undermining the prospects of a physicalist or naturalist worldview to develop an interest in a theory of intentionality. The sciences of the mind are replete with seemingly intentional terms: sensations, perceptions, memories, or plans, to mention just a few. Even if one takes for granted that ultimately these items can be localized in a physical or a functional system (e.g., as brain processes, or states of a computational system), there remains a question to be answered about how these processes or states manage to be about things outside of the brain or the computational system.

9.3.1 Intentionality: Real and "As-If"

Indeed, while the concept of intentionality has been a focus of philosophical thinking about the mind for a long time, the primary reason intentionality came to its prominent place in recent philosophy of mind was due to the deep entrenchment of mainstream philosophy of mind in the broader cognitive science project.

As seen in the previous chapter, the original cognitive science was committed to three core ideas: internalism, computationalism, and representationalism. Cognitive scientists and allied philosophers endorsed the idea that cognition essentially consisted of representations being computationally operated on. Such representations where thought to be located inside the brain, or possibly a comparable cognitive system in a nonbiological cognizer. Representations were thought of as physical entities: brain processes in humans, comparable to information or data inside a computer. But theorists of cognitivist persuasion realized that this physical way of looking at representations—as physical things involved in cognitive operations—doesn't exhaust what it is to be a representation.

Consider words or sentences. Typically, these are written down or spoken. They are physically present as visible symbols or acoustical perturbations. Yet, characterizing words as physical structures misses the fact that words and sentences have meanings. When I say, "It is raining here," I am not merely producing waves in the air; I am saying something meaningful about the weather in my environment. The sentence I spoke displays intentionality or aboutness.

Many thinkers attracted to cognitivism saw this intentionality or aboutness as present not only in language, but in representation in general. Cognitivists hold that there are representations outside of language, for example in the brain or in the visual system—remember Marr.

Just like words and sentences, representations then have two sides: a physical shape and intentionality or aboutness. So, when a cognitive scientist points at a process in the brain as a representation of a three-dimensional object, they aim at something double-faced. In fact, just as one can describe the words purely as geometrical shapes, one can characterize the cognitive scientist's neural representation by mentioning just its neural properties without mentioning what the representation is supposed to be about. The neural characterization will describe, for example, certain neurons in certain parts of the brain that are activated, in which electrical and chemical processes happen. But the process is also supposed to have intentional properties: It is about what it represents. The question now arises how the neural properties relate to the intentional properties. When raising a problem for physicalism regarding phenomenal properties, one could ask: How and why do these and these neural processes lead to or have a phenomenal feel? In the same way one can ask about intentionality: Why do these neural properties lead to aboutness? What stretches the bow of intentionality? Analogous questions can also be posed when the cognitive system is a computer or a robot.

In short, the problem of intentionality, or the problem of aboutness for classical cognitive science, is how representations inside the cognitive system manage to represent whatever they are about. Without an answer, representation, a core notion of cognitive science and its allied philosophy, becomes unfounded.

But perhaps the worries are exaggerated. Perhaps intentionality or aboutness flows forth from computational processing. When a computational system operates on data, inputs, or symbols, intentionality comes about naturally. On such a view, a well-programmed computer or robot would, just by running its programs, display intentionality. The idea that computationalism could account for intentionality so easily was attacked in a famous argument that became known as "the Chinese room argument" (Searle, 1980). The Chinese room argument was directed against those who believed that the kind of artificial intelligence systems built by the first generations of cognitive scientists were genuinely intelligent, and not merely acting as if they were intelligent. The artificial intelligence systems targeted in the Chinese room argument worked according to the classical cognitivist recipe. They received information or representations as input. Then they performed computational operations on this input to generate new representations. Which operations were performed was prescribed by the program of the system, which can be thought of as a set of rules of the form "if this condition occurs, then do that." Some representations generated along the way might be further operated upon, but eventually some output representations are generated. Searle claimed that a system built along these lines would never acquire intentionality. Therefore, such systems don't mean anything by their outputs, and it would be mistaken to ascribe intentionality or intelligence to them.

The Chinese room argument pivots around a thought experiment. Searle asks to consider a person in a room with input and output shutters. Through the input shutter the person receives strings of symbols. These are meaningful questions in Chinese. Crucially, the person in the room doesn't have any level of mastery of the Chinese language. Yet, the person has access to long lists of rules which prescribe how to operate on incoming strings of symbols, to change them into novel strings of symbols. These then are sent to the output shutter. Perhaps it would be possible to construct a rule book that made it look—to an outside observer—as if the person gives the right answers to these questions. The outside observer might be inclined to think the person in the room did understand Chinese and provided intelligent answers to what was asked. Yet, this would be mistaken, Searle insists, for the person only follows rules on what are, for that person, utterly meaningless symbols. The person just manipulates shapes in a regulated way, nothing more. The Chinese room argument extends this assessment of the human individual to any program built according to the symbol-manipulation blueprint. In this way, it does not only affect artificial intelligence, but it raises the problem of intentionality for cognitive systems as treated by classical cognitive science. If this is right, such systems don't have intentionality. Therefore, if such models are meant to characterize human cognition, they miss the crucial aspect of aboutness.

The Chinese room argument has generated heated discussion—a first batch of commentaries was published as part of Searle (1980). Many rose to the defense of cognitivism. A very common reply was to point out that the situation in the Chinese room provided only a fragment of the cognitivist picture of cognition. Any symbol-manipulating system in a genuine cognizer would be connected to many other cognitive systems, such as perceptual systems, memory systems, and control systems. Only when embedded in a wider operational whole would a symbol-manipulating system acquire intentionality. Some argued that the wider system should include action systems, so that the minimal unit for cognition-with-intentionality should be capable of action—it then could be an organism or a robot.

Another critique of the reasoning in the Chinese room argument took it as valid with respect to symbolic, but not to connectionist-style models of cognition (Churchland & Churchland, 1990). In the Churchlands' line of reasoning, symbolic models of cognition, such as the language of thought hypothesis, misconstrue cognition as manipulation of symbols. Thinking of cognition as symbol crunching misses the mark because cognition is a matter of the dynamics of neural networks. The thought experiment on which the Chinese room argument is built is appealing only when it is assumed that the man in the room is a symbolic system. The conclusion that such a symbolic system has no intentionality is correct. But if the argument would model the person in the room in a more accurate, neural-dynamics-based way, the conclusion that the system doesn't have intentionality would no longer be obvious. After all, is our brain in some sense not like the person in the room? And surely, from the neurocentric point of view defended by the Churchlands, it is the neurocomputational dynamics of the brain that makes us intentional.

9.3.2 Intentionality, Representation, and Intensionality-with-an-S

Both the Chinese room argument and many of the replies to it do not rely on a theory of intentionality or representation. Instead, intuitive notions of meaning, intentionality, and understanding are invoked to support Searle's claims or replies to it. Some philosophers took a different, more technical approach to dealing with intentionality and representation. These philosophers explicitly aimed at a theory of intentionality, rather than targeting artificial intelligence. Such a theory of intentionality would deal with the general problem of showing how intentionality could have a place in the natural world. It, thus, would provide a naturalistic, science-respecting response to Franz Brentano's claim that the mind's intentionality proved that physicalism was incomplete. Sometimes the project of formulating a theory of intentionality that philosophers of mind took upon themselves in the latter part of the twentieth century is therefore called a project of "naturalization" of intentionality (as implicit in, e.g., the title of the landmark publication (Millikan, 1984).

Crucially, philosophers engaged in the project of naturalizing tended to see intentionality as representation. For something to show intentionality was to be about something else, and for something to be about something else was to represent it. Such equation of intentionality and representation can be easily illustrated in the case of language. The sentence "It rains here" can be easily thought of as being about the weather in the environment by representing the weather. But philosophical theories of intentionality aimed for more than the intentionality of language. Intentionality was thought of as something characteristic not only of language but of cognitive processes in general. Cognitive processes can occur in the visual system in the brain, for example. These cognitive processes display intentionality, too: They concern the seen world. If intentionality is equated with representation, the aboutness of processes in the visual system implies these processes are representational. Many philosophers made that assumption. Doing so was very natural in the context of cognitive science, where representations are seen as ubiquitous in cognitive systems.

But is the equation of intentionality and representation unavoidable? It is not if there are forms of intentionality that are not representational. To see that this is possible, consider pain. When you have a headache, your headache seems to be intentional, about your head, but does it represent your head? At least it doesn't seem impossible to hold that your experience is intentional, but not representational.

Perhaps it is hard to get clear about the relation between intentionality and representation because it has not been made very clear what representation is. How can one answer questions about representation without knowing what "representation" means? This is a genuine concern. But it can be met. By the time philosophers of mind began proposing theories for naturalizing intentionality, a specific notion of representation had gained a firm foothold in philosophy. It came in the wake of how the logician Gottlob Frege (1848–1925) had analyzed the meaning of descriptive sentences in natural languages. Frege proposed thinking about the meaning of sentences in terms of what makes them true or false. What makes a declarative sentence true are its *truth conditions*. The truth condition of a declarative sentence are those situations in which it is true. "It rains today" can be true under a variety of conditions—all these being situations in which rain falls somewhere today. A declarative sentence is false when its truth conditions do not obtain. "It rains today" is false in all those cases in which it remains dry all day.

Philosophers aiming at naturalizing intentionality modeled a general notion of representation on this analysis (Evans, 1982). This more general notion was supposed to apply to various ways of representing, or various formats of representation. These could be linguistic, as for the original descriptive sentences, but there could equally be representations of a different kind: neural or pictorial, for example. Still, what was supposed to be common to all these representations was that they, in some way, specified, pictured, or described something and could be true or false in doing so. So, a picture might represent that Marilyn Monroe had blond curly hair, and if it did, it would be a true representation of her. A picture of Marilyn Monroe with black stiff hair would be false. Note that if one construes the notion of representation by means of truth conditions, representations must be able to fail or be false. For something to be a representation of something else, it must be possible that the representation misrepresents. A picture of Marilyn Monroe can misrepresent her, but Marilyn Monroe herself cannot. Marilyn can't be other than, or misaligned with, Marilyn, in the way that a picture can. Therefore, a picture represents Marilyn, but she herself doesn't. Similarly, her shadow can't misrepresent her. It just goes where her body goes. Depending on the light, it might disappear, but otherwise, it will always follow Marilyn. It is always aligned to her. Representing requires that there is a distance between the representation and what is represented, such that the representation is either aligned or not aligned to what it is about—to its truth conditions. Yet another way to state this is that representation involves a norm, according to which it can be judged. That norm is truth—or something very similar, like "veridicality" or "accuracy" or "correctness." A representation lives up to this standard if its truth conditions obtain; it doesn't when its truth conditions don't obtain. Representation, understood in terms of truth conditions, is "truth evaluable."

Truth evaluability is a mark of meaningfulness. For a representation to meaningfully represent was analyzed as for it to be truth-evaluable. Moreover, the meaning of representations was often called the "content" of representations. Theories of the meaning of representations were called "theories of content." As intentionality was equated with representation, theories of content were taken to be theories of intentionality.

One further distinction made by Frege needs attention here. It is the distinction between sense and reference that was already mentioned in chapter 5 (section 5.2). Frege famously distinguished between the *reference* of a linguistic expression and its *sense* (Frege, 1892/1948). Reference is what is described, sense how it is described. Consider Emily Brontë. It is clear what, or rather who, is meant by the name: the historical person Emily Brontë. But the same person could also be described in different ways: as the writer of *Wuthering Heights*, or as the second youngest Brontë sibling. "Being the author of *Wuthering Heights*" and "being the second youngest Brontë sibling" have different senses. It is not so that if one knows the sense of one of these, one knows the other (see chapter 5). Therefore, someone can know that "This is the house where the author of *Wuthering Heights* lives" is true, without knowing that "This is the house where the second youngest Brontë sibling lived" is true. This example shows the importance of recognizing sense over and above reference. If one considers only reference, the sentence "This is the house where the author of *Wuthering Heights* lives" and the sentence "This is the house where the second youngest Brontë sibling lived" are equivalent. But the example invoking knowledge shows that sense matters. It matters, sometimes, how something is represented. A final complication: when sense matters, as in the example invoking knowledge, one sometimes speaks of intensionality (with an "s"). In intensional-with-an-s phenomena, it does not just matter that one is intentionally related to something, but it also matters how what one is related to is represented. Following Frege's analysis, it is often assumed that all representation involves a way of representation—sense over and above reference. Any phenomenon that is representational then is intensional, over and above being intentional. If the visual system involves representations, these represent-in-a-way and display intensionality.

So, when philosophers started constructing theories of intentionality or "theories of content" to account for representation both in and outside language, in tune with cognitive science, there was supposed to be a very tight link among intentionality, meaning, representation, and content. Indeed, globally, these notions were understood in terms of each other. All forms of intentionality were understood as representational. And representation was understood as implying truth-evaluability, the possibility of misrepresentation, and a way of representing or intensionality. Notice then, that all intentionality was modeled on representation in language.

Representations were assumed to be present both at what philosophers called the "personal" and the "subpersonal level" (Dennett, 1969). Suppose you visually perceive a dog. In the standard analysis, by just having that visual experience, you represent the dog. The experience is personal-level in that it belongs to, or occurs to, you as a person. It is you who perceive. Therefore, your personal-level experience represents the dog. But at the same time, in line with the broader ideas animating the cognitive sciences, representations are supposed to occur in the cognitive system(s) that are causally involved in our experience. Before you become aware of the dog in your experience, a lot of information processing or computation has happened inside your visual system to make your experience possible—think about Helmholtz, Marr, or perhaps neural networks. Representations allegedly have been produced during those processes. Yet they remained below your personal radar, in that you were never aware of them. They are like the production processes that lead to the consumer goods you buy. They are necessary for the product to exist, but you need not be aware of them to enjoy or use the product. They happen subpersonally.

9.3.3 Theories of Content

The emergence of the project of accounting for representation wasn't entirely new, nor was the equation of representation, meaning, aboutness, and intentionality. For representationalism about the mind had been around for a long time, as we have seen in the previous chapters. Locke and Hume held opinions about what made their mental entities—ideas—intentional or meaningful. They believed that ideas were about other ideas, about impressions or about the world (certainly so in Locke) because of a resemblance relation.

Locke and Hume thought about ideas—or at least some of them—in terms of mental images. Seeing and remembering were taken as involving ideas that were like images or pictures. Now, how does a nonmental picture, for example, a portrait, represent? By resembling what it pictures, that might seem quite obvious. To the extent that ideas are picture-like, it is natural to think about them too as having aboutness because of resembling something.

The resemblance theory of intentionality got bad press in the decades and centuries that followed. One particularly impactful attack was delivered by Nelson Goodman (1968). Goodman pointed out, inter alia, that resemblance was symmetric: If A resembles B, B resembles A. But representation or intentionality is asymmetric: My thought about a dog is about the dog, but the dog is not about my thought. Further, resemblance is a weak notion. Everything resembles everything else in *some* respect. The same picture can resemble various situations. Wittgenstein gave the following example: "I see a picture; it represents an old man walking up a steep path leaning on a stick.—How? Might it not have looked just the same if he had been sliding downhill in that position? Perhaps a Martian would describe the picture so" (Wittgenstein 1953, § 60).

Like Goodman, Wittgenstein invoked the open nature of resemblance as requiring constraints on when we actually use pictures in meaningful ways. We do so all the time. Think about traffic signs or icons on a laptop screen. That these seldom lead to confusion, despite in principle standing open to many interpretations, just means that additional constraints are at play. Goodman famously argued that these constraints are conventional, illustrating his point through the diversity of artistic styles (Goodman, 1968). Just like we have the convention to use red as a stop sign, art has conventions for picturing things in certain ways, and these conventions can change—think about the difference between a painting in linear perspective and a Cubist canvas.

So, it should come as no surprise that when theories of intentionality were placed high on the philosophical research agenda, somewhere in the 1970s, philosophers avoided relying on resemblance. A notion that appealed much more to them was causation (Stampe, 1977). In fact, there is a practice of speaking about things that are stably causally related in terms of meaning: Smoke means fire. Paul Grice (1913–1988) and later Fred Dretske (1932–2013) called meaning based on such relations "natural meanings" (Dretske, 1986, 1988; Grice, 1957). The aim of causal theories of content was to build a theory of intentionality for representations—in the brain or elsewhere—on such a causal basis. Effects would represent their causes. Proponents of causal theories of meaning pointed out that causality compared favorably to resemblance. For causation is asymmetric. One cannot, in general, invert the arrow of causality. Causes cause their effects, but, overall, effects don't cause their causes. Further, grounding intentionality in causality nicely accommodated the goal of providing a naturalistic, scientifically respectable theory of content.

However, even if causality doesn't have some of the problems of resemblance, it too is a problematic candidate for carrying all the weight of a theory of content and intentionality. To begin with, just like resemblance, causality applies very broadly. Causality is ubiquitous. But meaning, representation, and intentionality aren't. I cause my shadow, but my shadow doesn't represent me, at least not in the sense of specifying me in a truth-evaluable way. Causality lacks the normativity that representation has. A way to illustrate that is to think about a visual representation of horses. To the question: "Why is the horse representation in your visual system a representation of a horse?" the causal theorist will answer that it is because it is caused by a horse. But now suppose that, on a dark night, you bring your visual system to a dimly lit cow (Fodor, 1990). Perhaps, under such conditions, the visual system might still produce a horse representation. But if causality is all that is at play, the causal theorist is obliged to conclude that the meaning of the representation is no longer horse, but horse or cow on a dark night—because both are causes of the representation. This problem is called the "disjunction problem" because a causal theory must include all the actual causes into the meaning of a representation, ending up with a construction of the meaning as "this cause or that cause or yet another cause or . . ."

Perhaps the way out is to add something to causation. Consider a traditional thermostat (Dretske, 1988, pp. 86–87). An electric circuit runs through the thermostat, and when electricity flows through, the heater is on. But one part of the circuit in the thermostat consists of a bimetallic strip—a strip made of two different metals. Like all other metals, the strip will expand when the temperature in its surroundings rises. However, one part of the strip will, because of the kind of metal it is made of, expand more than the other. This will make the whole strip bend and eventually the circuit will be interrupted, and the heating will stop. Once the temperature starts going down, exactly the reverse will happen.

Perhaps the bimetallic strip could be bended by other causes than the room temperature—think about heating it up by means of a laser beam. Still, one would want to say that the bimetallic strip has a unique relation to the temperature, and not to laser beams. The reason for this is its function: The function of the bimetallic strip is to register the temperature so that the overall function of the thermostat can be performed: keeping the room at a constant temperature. The bimetallic strip is chosen for this because it has a causal connection to the room temperature, manifest in its expansion. Many philosophers think the situation sketched justifies describing it by saying that the strip has the function of informing the thermostat, and by this, the whole heating system, about the ambient temperature. So, theorists added function to causality to come to a better theory of intentionality (Dretske, 1986, 1988). According to this new approach, the bimetallic strip is about temperature. This approach claims advantages over both resemblance and causation-only theories. It retains the asymmetry from causal approaches, but it constrains causality. It seems to have resources to deal with the disjunction problem, by its reliance on function. Our bimetallic strip is about ambient temperature, not about laser beams, even if these could cause it to expand too, because its function in the whole system is to convey information about heating.

So far so good, perhaps, but functions of thermostats are designed by thinking humans. Humans and their thoughts are intentional. So, if one invokes human-designed functions in a theory of intentionality, one seems to make intentionality depend on preexisting intentionality. This makes the approach circular, and still does not provide a naturalistic, scientifically respectable account of intentionality.

In an extremely impactful series of publications, Ruth Millikan provided a theory of intentionality that invoked evolution, not preexisting intentionality, as the source of functionality in biological systems (Millikan, 1984, 1993). A prominent feature of the theory of evolution by natural selection is that it allows to account for biological function in selective, nonintentional terms, as noted in chapter 4. The genetic makeup of organisms changes by accidental rather than premeditated causes, and those changes that lead to better functioning-in-an-environment are retained in the selective process. In this way natural selection gradually leads to well-adapted bodies and well-adapted organs. But selection applies to more than somatic structures. One of Millikan's prominent examples concerns the dances of bees to indicate the location of nectar. When they have found nectar and moved closer to their peers, bees start moving in certain ways that correlate with where the nectar is. Their companion bees can, by watching them, find their way to the nectar without first having to perceive the nectar themselves. On Millikan's analysis, the dance is an evolved representation, because it has a systematic—ultimately causal—relation to the location of the nectar that can be consumed by companion bees. There is systematic correlation on the production side, which is exploited on a consumer side, and therefore the system is representational and intentional. The dance has truth conditions: there actually being nectar at the place indicated by it.

Millikan's theory of content is known as "teleosemantics." It is "semantics" because it is meant to be a theory of intentionality, content, and meaning. The "teleo" is in there because the theory is teleological: It relies on functions with goals—the ancient Greek word for goal is "telos." Further, Millikan's theory is called "aeteological," because according to Millikan's theory functions are determined by evolutionary history. The heart makes noise when it beats to circulate the blood through the body's veins. But making a beating sound is not what the heart is selected *for*. If it just made the sound, the heart would not have been retained in the process of evolution. The heart was selected for pumping the blood. That's what gave an advantage to organisms that had one. So what the function of the heart is, is determined by evolutionary history: to pump, and not to make noise.

Teleosemantics is taken, by Millikan and many others to offer a tool to analyze representation in the brain. Some neural systems are producers of representations for other neural systems that act as consumers. Neural processes might occur in the perceptual system of an organism whenever the sensory receptors are activated in certain ways, and this activation might cause further activity in, for example, action-preparing neural systems. Complex systematic correspondences between the world and producer systems—think bee dances in the brain, and beyond—and between producer systems and consumers systems might have evolved—think bee reactions to dances and beyond. Suppose there is a visual producer system for tracking certain objects, for example, flies in frogs or spiders in humans. Such systems have evolved because the objects they detect are important to the organism. Moreover, it is safe to assume they are coupled to consumer systems so that they are relevant for action. Such producer systems are never foolproof. Occasionally, they might be activated by something that looks very much like a fly, by a spider, by a knock on the head, or, who knows, by a laser beam. Because of their physical makeup, producers are disposed to react in the same way as when confronted with what matters evolutionary. But, by referring to what matters to them, and what they were selected for, a distinction can be made between what they are merely disposed to react to, and supposed to react to. This is the way in which normativity has a place in teleosemantics. Normativity is a consequence of evolutionary function.

Jerry Fodor (1990) argued that teleosemantics didn't work. In his view, it missed a central aspect of meaning in not allowing for intensionality-with-an-s. It missed sense, or way of representing. Therefore, it did not give a theory of representation. Therefore, it did not offer a theory of intentionality. And, therefore, it failed to naturalize intentionality.

Fodor illustrated his argument by appealing to a mechanism in the frog's brain for detecting flies. Teleosemanticists propose to construe such system as a representation producer because of its links to fly-catching processes in the frog's brain and body. Yet, they don't have a right to do so, according to Fodor, because teleosemantics remains silent on the way the fly representation represents the fly. Remember, when we speak about Emily Brontë, we can do so in many ways. Emily can be the only one person targeted, but the targeting can be done in several ways. There is a referential aspect to meaning—involving here a unique individual—and an intensional aspect: the way in which the referred to person is characterized. Several different intensional meanings converge on one referential object. Fodor criticizes teleosemantics for not giving a clue about the intensional aspect of meaning. Evolution determines that the frog has fly detectors, but evolution doesn't determine that these should be thought of as detectors of flies, rather than, for example of "little nutritive black flying shapes that were present in the frog's evolutionary history," or any other description that targets flies.

If Fodor's argument is taken to refute Millikan's approach as a theory of intentionality, the usual equation of intentionality with representation is upheld. Fodor's argues that teleosemantics offers no insight into intensionality-with-an-s, and therefore is deficient as an account of representation. To say that it therefore is also deficient as a theory of intentionality is to accept the equivalence of representation and intentionality. As said, this equivalence is standardly taken for granted. But it needn't be. If intentionality is not equivalent to representation, there could be intentionality without representation. Millikan's theory could be fine for intentionality, but not for representation.

Hutto and Myin (2013) have raised general doubts about equating intentionality with representation. They defend that one can have intentionality without representation. For instance, perception can be intentional without being representational. To see this point, contrast seeing an apple in front of one, with judging "There is an apple in front of me." The judgment contains a description of a fact, and expresses that the speaker asserts the truth of that description. The description characterizes the apple by means of the concept "apple." If perception is contentful, or representational in a truth-conditional sense, it is like the judgment. But, so Hutto and Myin claim, perception is not contentful in that sense. Perceiving an apple in front of one is not some perceptual form of judgment. Perception doesn't consist of the production of something like descriptions. Perception is specific—seeing an apple is different from seeing a waterfall. But perception doesn't specify: It doesn't picture, characterize, or describe what one perceives. How then, if not by containing a description, is perception of an apple intentionally linked to the apple? Because of perception's evolutionary history. Evolution has linked perception and what is perceived, without using descriptions as a glue.

In short, Hutto and Myin are concerned that notions such as "having truth conditions" or "being a representation" are sometimes ascribed without license. No good reasons have been given, they claim, for the very common assumption that perception is always representational, and the same goes for many other cognitive phenomena. Equally, no good reason exists to assume that biological normativity as in bee dances, or in brain systems involves truth conditions.

According to Hutto and Myin, if one assumes that all intentionality is contentful, that is, involves truth conditions, one ascribes properties that are characteristic for evolutionary late-coming capacities, such as language, to basic capacities. But then one takes representing—in the sense of having truth-conditional content—as primitive. Instead, one should recognize that basic capacities like perception are not inherently representational, and provide an account of how truth conditions came to exist via an account of how practices of truth telling emerged.

Hutto and Myin's analysis rejects a representational view of the mind that has been hugely influential in thinking about the mind. As emphasized throughout this book, representationalism was endorsed by philosophers in the tradition of what Reid termed "the way of ideas" and gained a firm foothold in scientific approaches to the mind, from psychophysics to cognitive science. By opposing it, Hutto and Myin join a tradition including many of the thinkers that were discussed in chapters 3, 4, and 6. In chapters 10 and 11, we'll see how this tradition is revived in contemporary thinking about the mind.

Bibliographical Note for Chapter 9

The literature on consciousness is enormous. One strand that I have left out is *illusionism* about consciousness. The core idea of illusionism is that people are mistaken about consciousnes. People ascribe to consciousness properties it in reality doesn't have, such as having special phenomenal qualities. The philosophical problem of consciousness then is not to give an account of how there can be phenomenal properties, but rather to account for why we wrongly think there are phenomenal properties. In reality there just are physical processes with physical properties. Illusionism is often traced back to Daniel Dennett, and several of his publications, including *Consciousness Explained* (Dennett, 1991a). Dennett's illusionism is sometimes taken to be about eliminating consciousness, rendering his position extremely radical. On a different, and I think more accurate, construal, Dennett's position is not that consciousness should be eliminated, but rather that the philosophical notion of "qualia" should be (Dennett, 1991a, chapter 12). Illusionism has been further developed by Keith Frankish (Frankish, 2016).

The Chinese room argument has also generated an enormous amount of responses and further reflections. Many, though certainly not all, trends in the later literature can be discerned in the early batch of responses that were published together with Searle's original article (for which, see Searle, 1980).

The widespread analysis of perception as representational and contentful—in the sense of ascribing properties—has met more challengers than presented in this chapter. One important philosophical strand of opposition can be found of the work of Charles Travis and Bill Brewer—often labeled as "direct realism" (Brewer, 2011; Travis, 2013). Travis insists that "the senses are silent." Or, perception is not conceptual: it doesn't represent the world in terms of generalities, like language does. Instead, perception gets us in touch with particular, what Travis calls "historical," things (Travis, 2013).

10 Beyond the Head

> The skin is not that important as a boundary.
> —Burrhus Skinner (1963, p. 953)

> Ask not what's inside your head, but what your head is inside of.
> —William Mace (1977, p. 43)

10.1 Embodied Cognition in Psychology and Linguistics

The philosophers discussed in chapter 6 stressed the importance of paying due attention to what Wittgenstein called "forms of life" when it came to understanding mental phenomena. They thought that without linking perceiving, intending, and thinking, for example, to their natural, biological, and sociocultural surroundings, only distorted pictures of and mistaken conclusions about these phenomena would be obtained.

But the cognitivist and functionalist views of the mind that became prominent in the sciences and philosophy in many ways led to an opposite tendency to disregard or pay less than full attention to everything apart from the brain when it comes to understanding cognition. Several strands in the cognitivist view promote such an attitude.

If cognition is seen as computation over representations of the environment, the environment counts only insofar as it is represented. Moreover, if the computations are assumed to occur in the brain—a common assumption—the body plays only a supporting role in cognition. With the thesis of the multiple realizability of cognition, even the brain was demoted and given a secondary status. What mattered was not the brain as the specific material organ it is, but its capacity to engage in abstractly specified computational activities—activities that allegedly could be carried out by systems of quite a different material makeup than brains as we know them.

Hubert Dreyfus's philosophical critique of contemporary artificial intelligence highlighted the neglect of the body and the environment in theories of cognition. A different line of drawing attention to the body emerged through scientific research. Increasingly people operating within a broadly cognitive science framework started claiming an important and unexpected role for the body in cognition.

Some of this work was straightforwardly empirical. In one study, for example, people were shown a sequence of nonsense syllables while assuming a specific body posture (Rand & Wapner, 1967). It was found that they recalled the syllables better when they assumed the same body posture they had during the initial exposure to the syllables. A later paper (Riskind, 1983) reports that it has been found that people more easily remember pleasant events from their own history—their autobiographical past—when they take a posture and facial expression that expresses positivity. So, they remember positive events better from an erect position with a smile on their face. Unpleased events were better recalled with the shoulders hanging down and an unhappy face (Riskind, 1983).

Dijkstra et al. (2007) note that these results can be framed in an "embodied" approach to memory, according to which remembering an event is a kind of simulation of the original event, and so includes reliving perceptual, motor, and affective aspects of the original experience.

Very similar ideas have been proposed in embodied approaches to semantics—the science of linguistic meaning. One strand of embodied semantics is guided by the idea that the understanding of the meaning of a sentence derives from a simulation of experiences and activities which are described in, or related to, the sentence (Barsalou, 1999; Glenberg & Kaschak, 2002; Kaschak et al., 2014). In understanding language about objects, one is supposed to simulate an experience that one would have if one would be perceiving these objects; in understanding activities one is supposed to simulate carrying out these activities. Thus, one understands "John was hammering a nail in a piece of wood" by imaginatively simulating what would be seen and heard when that John was hammering, by what movements one would have to make to hammer and so on.

Embodied semantics points to neuroscientific evidence for support. For example, Hauk et al. (2004) found that motor areas in the brain associated with the hand, foot, and mouth show activation when the words "pick," "kick," and "lick" are heard. Further evidence for the hypothesis has been claimed to derive from experiments that show that hearing particular sentences facilitates movements which are congruent with the sentence's meanings (see Kaschak et al., 2014). For example, Glenberg and Kaschak (2002) found that participants in an experiment were faster in making movements with their arms toward the body when hearing a sentence that says they are given a pen, as opposed to making movements away from the body.

Embodiment also plays a central role in the research on the metaphorical nature of cognition by Mark Johnson and George Lakoff. In their impactful books *Metaphors We Live By* and *Philosophy in the Flesh,* Lakoff and Johnson (1980, 1999) argue that the ways we perceive, act, and think is structured by metaphors based on the embodied ways we relate to our surroundings. For example, our bodily makeup not only determines our orientation, our ways of moving and perceiving in space, but also the ways in which we talk about space in language, as well as how we talk and think about nonspatial matters. The words "up," "down," and "in front of," are those of creatures with our ways of being, acting, and perceiving in our spatial surroundings. Moreover, this basic and bodily grounded way of conceptualizing space is thought to form the groundwork for other uses of speaking and thinking which do not literally involve space, such as when one metaphorically talks of being "*on top of* the situation" or says that "the meaning *is right there in* the words." Lakoff and Johnson invoke what they call "image schemas" to account for this pervasive influence of the body on cognition. Such image schemas are held to be cognitive neural structures (Lakoff & Johnson, 1999, p. 20), that are built up as a consequence of embodied interaction, and that play a role in organizing experience and thought. Body-related image schemas are the basis for more abstract schemas, which are found in all forms of language, including the language of mathematics (Lakoff & Núñez, 2000). In this way, instead of being grounded in some abstract world having nothing to do with earthly, concrete matters, mathematics becomes grounded in the concrete environment and the way humans interact with it through embodied activity.

How far do the embodied views discussed so far take us from traditional cognitivism? They all draw attention to the role of the body in cognition. Especially in embodied semantics, the idea is promoted that cognition is based in bodily activity. But insofar as this role of the body can be mediated by neural representations of the body, the embodied approaches to memory and semantics surveyed here can stay in line with the main tenets of cognitivism and retain the idea that cognition is computational, representational, and internal. In various places, practitioners of embodied linguistics have argued that they hold different views of representation than standard cognitivists. Barsalou (1999) has presented his view as disproving the idea that representations are amodal, or abstract and free of sensory elements. Against the view that the mental concept of "dog" is amodal, Barsalou holds that the concept involves sensory aspects, such as how dogs perceptually appear. Lakoff and Johnson have distanced themselves from representationalism, or in any case, standard representationalism. They have insisted that the body-based concepts that structure cognition, even if they are called "representations" are "flexible pattern [s] of organism-environment interactions" (Johnson & Lakoff, 2002, pp. 249–250).

10.2 Embeddedness and the Extended Mind

If some theorists started emphasizing embodiment, others did the same for embeddedness (see Shapiro & Spaulding, 2024, for an overview). Cognitive agents are not only embodied, they also act in environments. Which environment one is in might matter for what one is cognitively capable of. Moreover, the environment might play a role that one can think of as being "cognitive." Hence, the internalist view broadly endorsed by cognitivist theorists since the cognitive revolution (chapter 7) might have to be revised. That is, cognition might not only be happening in the head, but the locus of cognition might include the environment.

To warm up to the idea, consider the philosopher John Haugeland's example of finding one's way to San José (Haugeland, 2000, p. 234). Haugeland notes one can think of being able to find one's way to San José in terms of a "mental map"—knowledge of where to start, where to change direction at places in the itinerary, and so on. But one will manage to get to San José without such knowledge. All one has to do is enter the highway to San José at an appropriate point and stay on that highway. The knowledge, so Haugeland suggests, is jointly in the environment and the person, instead of being just in the person or their head. The road itself is an integral part of the capacity to go to San José.

The idea that parts of the environment can become part of cognition has become known as "the extended mind thesis." It was proposed by Andy

Clark, originally together with David Chalmers, in their iconic paper, "The Extended Mind" (Clark & Chalmers, 1998). Clark and Chalmers argued that internalism about cognitive processes is untenable. Instead of being confined to the brain, or even the body, the mind, so they claim, sometimes is extended, in the sense that things belonging to the environment become part of cognition. At least in some cases, cognitive processing consists of operations in the environment outside of the body. Clark and Chalmers refer to Kirsh and Maglio (1994) for empirical support. Kirsh and Maglio (1994) found that players of the game *Tetris*, which involves rotating figures made up of small squares, sometimes engage in rotating objects on the screen instead of imagining the rotation in the mind's eye. Clark and Chalmers claim that the player's physical action of causing the on-screen rotation is a cognitive process in its own right, just as an imagined rotation. Invoking a "parity principle," they claim that what happens outside of the head should be treated equally to what happens inside the head. Clark and Chalmers hold that in such cases as described in the *Tetris* example, the mind extends outside of the head and into the environment. To illustrate their point, Clark and Chalmers present the imaginary case of Inga versus Otto. Inga has a well-functioning memory and one of the things that she is able to remember when she needs it, is that the Museum of Modern Art in New York is located on 53rd Street. In contrast to Inga, Otto suffers from Alzheimer's disease, which causes him trouble to remember such facts as the address of the MoMA. Otto uses a notebook to cope with the situation. He always carries it with him, to note things he assesses as being particularly relevant. So, when Otto wants to visit the MoMA, he digs up the address from his notebook. In the end, relying on his booklet, Otto is able to get there, just as Inga is, relying on her natural memory. Anyone endorsing a functionalist picture of cognition, so Clark and Chalmers claim, should endorse the statement that the notebook is as much part of Otto's cognition as are the brain processes subserving Inga's memory. The notebook and its contents play the same functional role as what is contained in Inga's natural memory. Thus, at least in such cases, the mind extends into the environment. Instead of internalism about cognition, Clark and Chalmers advocate what they call "active externalism": Cognitive processing can involve the world in a way in which parts of the world play more than a causal role in cognition. Instead of merely triggering cognitive processes "from the outside," worldly entities can become steering and *constitutive* parts of cognition.

In further work, Andy Clark has elaborated and defended the extended mind thesis (Clark, 2003, 2008). He has pointed out that extension goes beyond cases suggested by the parity principle and the Otto/Inga example, in that the role of the environment might go beyond replacing or compensating for lost natural cognition. Clark paints a picture in which the mind gets transformed by incorporating external entities. Such incorporation might allow minded organisms to perform cognitive tasks that would be impossible without those external objects (Clark, 2003). Think about how pen and paper allow people to make more and more complex calculations. Or about how repeated cycles of sketching on a sketchpad and assessing what's drawn allows artists to make compositions impossible to figure out "in the head" (Clark, 2003, p. 76).

Clark identifies language as the most significant example of how the involvement of the environment leads to an extension of our cognitive capacities. In Clark's analysis, language functions as a partly external, integrated supported system that enables the emergence of cognitive capacities that would otherwise be unattainable (Clark, 2003, chapter 3; Clark, 2014, chapter 8). When making these claims, Clark is thinking about language in its tangible, worldly aspects, such as the signs on the paper or screen you are now reading, or the sound of speech. These are concrete objects and processes in the external world.

According to Clark, an experiment with chimpanzees reported in Thompson et al. (1997) shows how the introduction of a simple form of external symbols leads to a remarkable extension of the animals' cognitive potential. To understand the experiment, consider three orders of difference and sameness. The first kind of difference and sameness, let's call it the zero-order difference or sameness, is the difference or sameness between two elements. The two elements shoe and banana are different; the two elements shoe and shoe are the same. To be sensitive to the zero-order relation of difference and sameness between objects, one just needs to be able to react differently to different objects.

The next, or first-order difference and sameness, is the difference or sameness between pairs of things, depending on whether the pair contains different or the same things. The first-order difference and sameness thus concerns the relation of difference or sameness between the elements of a pair. To be sensitive to the first-order difference, one would have to be able to give a specific reaction to any pair of different items, and a different specific reaction to any pair of the same items. Or, one would show reaction R1 to all the pairs AB, BC, AC and reaction R2 to all pairs AA, BB, CC. Finally, consider second-order difference and sameness. Second-order difference and sameness concerns pairs of pairs. Second-order difference depends on the first-order difference or sameness between each of the pairs. If the first-order relation between the pairs is the same, as in the pair of pairs (AB, BC) the second-order relation is sameness. In this case, the pair AB is an example of the first-order relation of difference. But so is the second pair, BC. Therefore, we have twice the first-order relation *difference,* and this gives second-order *sameness*. In contrast, the pair of pairs (AA, BC) is an example of the second-order relation of *difference*.

Thompson et al. (1997) investigated a group of chimpanzees trained to associate a specific physical object, such as a yellow triangle, with any pair of identical things, for example, two shoes or two bananas. At the same time, they learned to associate another shape, such as a red circle, with any pair of nonidentical things, such as a shoe and a banana, or a cup and a shoe. In other words, these animals were trained to associate the first-order relation with a specific shape.

The experimenters then found that the group of chimpanzees that had been trained with the plastic shapes were able to perform a task that involved sensitivity to the second-order relation of difference and sameness. No chimpanzee that was not exposed to the training could perform the second-order task successfully.

Andy Clark explains the result in terms of hypothetical mental processes (Clark, 2014, pp. 171–172). The trained chimpanzees would experience mental images of the shapes they had been trained with. For them, the higher-order task would become the lower-order task of assessing whether their mental images are the same or different. Note that the extra training, being exposed to physical forms in the environment, forms a necessary step in the trained monkey's cognitive achievement. Without this link, there is no chain that connects the more difficult task with the simpler one. In other words, the example shows how the use of worldly objects—the plastic forms—expands and transforms cognitive capacities. A new capability is formed, due to the involvement and incorporation of external elements. In this process, cognitive capacities get transformed, and achieve a level of functioning that was not attainable without the external elements.

Clark and other theorists have borrowed the term "scaffolding" from the Russian developmental psychologist Lev Vygotsky (1896–1934) to describe the role of external objects and structures in cognition. Vygotsky has investigated the role of the social environment in the cognitive development of children. He saw the role more skilled peers play in educating less skilled persons as comparable to a scaffold in construction works. Just like a scaffold is an external structure that allows one to build higher than without the scaffold, the involvement of peers allows learners to reach levels of skills unattainable by independent effort. Scaffolds can be discerned in institutions (schools, universities), in practices (teaching, devising a curriculum), or in material objects (books, sketchpads, computers).

Clark points at language as playing a particularly important role in allowing for scaffolding. Language, in his view, is a piece of cognitive technology (Clark, 2014, chapter 8). It is language that allows us specific ways of structuring our acting, thinking, and planning, and so enables us to build further scaffolds for augmenting our cognition. That language can play this role, so Clark points out, is to a large extent because language is *in* the world. Spoken and written words are objects and, therefore, can become objects of attention and cognitive activity themselves. Language literally gives shape to thought and, in this way, opens the door for further thinking that would be impossible without it.

Clark distinguishes two ways to think about the relation between thought and the complex structured language we have (Clark, 2014, pp. 174–175). One way is to conceive of the structure of language as reflecting the language-independent structure of thought. Thinking would naturally have a linguistic of protolinguistic format and then impresses this on language. This is the view espoused by Chomsky and Fodor (see chapter 7). The different view is to see structured thought as we humans have it as a product of language. Basic forms of thinking allowed for basic forms of language. Because of its objective aspects, basic language then allowed for more complex thought, which allowed for more complex language, and so on. The involvement of external structures—utterances and symbols—comes to play a crucial role in the explanation of thinking.

By opposing internalism, or the view that the mind is in the head, the extended mind thesis and Clark's view on cognition oppose one assumption that had been taken for granted in standard cognitivism. Yet, the extended mind thesis and Clark's wider views of cognition are compatible with the other core tenets of a cognitivist and functionalist outlook. Clark and Chalmers (1998) and Clark (2014) explicitly endorse functionalism, computationalism, and representationalism. The extended view of the mind espoused by Clark is complementary to standard cognitive science. It wants to *add* to the main framework, without calling for a radical overhaul. Other theorists that have stressed the role of the environment and of embodiment have taken a less conciliatory stance.

10.3 Radical Embodied Cognitive Science

10.3.1 Radical Is Relative

In the history of Western thought, there is a tradition that calls for more than adding to or locally adjusting the core ideas of the cognitivist picture. We have already seen this tradition at work in Reid, Dewey, Skinner, Wittgenstein, Anscombe, and Ryle. Typically, though with important differences, these thinkers all posit that mentality cannot be disconnected from whole organisms, active in environments. They warn for reification or for the conception of the mental in terms of internal and private objects and processes. Typically, they are critical of mental representations, as conceived of by cognitivists.

The fortunes of thinking along these lines have waxed and waned. It was not the thought of Reid, but that of Locke, that had the dominant impact on experimental psychology when it emerged. In the heydays of behaviorism, the balance shifted, until the cognitive revolution happened. In philosophy, periods in which the influence of pragmatism (including Dewey) and Wittgenstein was large were followed by cognitive science–inspired functionalism and a broad endorsement of the inner and of mental representation. What is mainstream and what is outside of it, is relative, however (Chemero, 2009, chapter 1). Forms of embodied thinking in science and philosophy critical of the core ideas of cognitivism have become known as "radical" because they emerged in a context in which cognitivism was dominant. Indeed, Tony Chemero uses the term "radical embodied cognitive science" for approaches to the mind that reject the standard assumptions of cognitivism, namely internalism, representationalism, and computationalism (Chemero, 2009, pp. 28–29, cites Clark, 1997, p. 148, as the origin of the term). Two strands in scientific thinking about the mind that Chemero singles out as forms of radical embodied cognitive science are ecological psychology and dynamical systems approaches to cognition.

10.3.2 Ecological Psychology

One important motivation for cognitivism has always been the idea of "the poverty of the stimulus" (the term was coined by Chomsky, 1980, p. 34), expressing that what the environment furnishes to the organism in terms of sensory or other stimulation is in some sense poor, insufficient, or ambiguous. The poverty of the stimulus argument has played an important role in Helmholtz's call for the need to invoke processes of unconscious inference to account for perception. In general, in the context of cognitivist thinking about perception, the stimulation received by the sensory system is considered impoverished, because it leaves open multiple interpretations. A large square at a distance can give rise to the same activation of the retina as a smaller square nearby. Cognitive systems in the brain must process this ambiguous information to form an unambiguous percept.

James J. Gibson (1904–1979) was a foremost perceptual theorist who had never been convinced by the poverty of the stimulus line of reasoning. He argued that standard approaches to perception had too quickly decided that sensation should be the starting point for the study of perception (Gibson, 1963, 1966, 1979/1986). As seen in chapter 3, the notion of sensation used by psychophysicists was based in physics. Sensations were often thought of as mental correlates of the magnitudes measured and studied by physicists. For Gibson, the study of perception should not start by imposing the categories of physics on animal perception. Sensory organs have not evolved to be natural versions of the measurement devices that have been developed in the course of the growth of physics. Organisms see colors and hear noises but are quite bad when it comes to registering exactly the precise wavelength of a light or a sound. It is possible to study, as structuralist psychophysics does, how subjects of experiments react exactly when a controlled stimulus varies in its wavelength value, but such exercise doesn't yield much insight into perception, so Gibson contends. And it is a serious mistake to think that what is experienced in such experiments are the atoms of perception. Rather than unearthing the pure basis of all perception, such experiments only offer a window into very specific kinds of perception in unusually impoverished conditions.

Gibson thinks perceptual science should rather look at the ways in which biological organisms inhabit and act in their natural environments. Perception should be studied in an *ecological* way. Gibson was keen to point out that perceivers are organisms, with needs, that live in specifically structured environments or niches. They need to discover food, stable paths to move on, safe places to shelter, and so on.

Gibson stresses that ecological environments contain multitudes of energy patterns that are shaped by what is present in those environments. At any place near a tree or a table, patterns of light are available that are specific to that tree or table, in the sense that only that tree or that table would form that pattern of reflected light. Gibson calls the patterns of light formed by the objects relative to a point of observation in the environment the "ambient optic array" (Gibson, 1966).

The example with which Reid's notion of visible figure was illustrated in chapter 3, can be reused (section 3.4.4). You can take a picture of the same person from many spots. Why is that so? Because the light that reflects from that person is available at all those places. Moreover, the light is structured; it is shaped by the physical form of the person, and though the details of the structure change from one picture-taking position to the other, the patterns of variation are systematic and therefore allow us to recognize the same person in pictures taken from many positions.

Using the terminology of his time, Gibson speaks of the optic array as providing information about the objects in the environment, meaning that there is a strict correspondence between patterns of light and objects. When an observer moves in the environment, a succession of patterns ensues—the pattern for a tree changes because the point of observation changes. But the change of patterns in the optic array is regular and patterned, too: It can be predicted how the pattern formed by the tree will change when the observer stands up or moves around.

Gibson emphasizes that the ambient optic array is rich, rather than impoverished. Patterns are specific to objects, and patterns of patterns created when an observer moves even more so. This can be illustrated by the ambiguous square example. If a perceiver is able to move sideways while looking at the square, the patterns in the optic array will change differently depending on whether the square is large and far away or small and close.

Importantly, seeing, says Gibson, should be conceived as being directly in touch with objects in the environment by detecting, picking up, or resonating with information available in the patterns of ambient energy. In his view, perception is of the tree in an unmediated way, because of sensitivity to the tree-specific pattern. So, Gibson holds that perception is what he calls "direct". Perception is not due to a mental tree-percept, generated by inferential processes inside the organism operating on information brought inside the nervous system. Further, Gibson stresses that perceiving is done by the whole organism, instead of happening inside its brain. Gibson commits to perception as inherently embodied and embedded. In this vein, he writes, "I shall suggest that natural vision depends on the eyes in the head on a body supported by the ground, the brain being only the central organ of a complete visual system" (Gibson, 1979/1986, p. 1).

Of course, there's a further question for Gibson and his followers about *which* environmental patterns organisms become sensitive to. Gibson very explicitly relates perceptual sensitivities to what organisms can do in their environments. Consider the following example. Gannets are birds that hunt for fish. When they spot a fish near the surface of the water, they dive down at great speed to snatch their prey from the water. However, they must heed to retract their wings in time. If their wings would hit the surface of the water at high speed, they would break. Gibsonian scientists have investigated what in the diving gannet's optic array is specific for the moment for retraction. In short, when a gannet dives, the optical information available to the gannet changes in a way that correlates with the remaining distance to the water. By resonating with this changing pattern, the gannet can retract its wings at exactly the right time. (Lee, 1980; Lee & Reddish, 1981).

Gibson used the word "affordance" to express the link between what organisms perceive and their actions (Gibson, 1966, 1979/1986). Things in the environment allow perceivers to engage in certain activities. A chair allows a person to sit, or it affords sitting to a person. A hole in the wall allows the mouse to escape from a cat and thus affords escaping to the mouse. Gibson held that affordances can be directly perceived. Affordances are not perceived by first perceiving forms or neutral objects and then making inferences to what can be done with them. Instead, organisms detect energy patterns in the environment that are specific for certain affordances. Pursuing this idea empirically, ecological theorists have investigated the perception of an affordance offered by stairs: climbability. People can only climb stairs if the steps are not too high, relative to the length of their legs. People with longer legs can climb stairs with somewhat higher steps, of course. In general, for a stair to be climbable, the height of the steps must be in a fixed proportion to the length of the legs. In a classical study, Bill Warren showed that people can perfectly perceive whether stairs are climbable or not. Just by looking at stairs, they can promptly assess whether they afford climbing to them. Their judgments of a stair's climbability are correct: They assess stairs with the right ratio as climbable, and those that deviate from that ratio as not climbable (Warren, 1984). Judgments differ, but remain accurate, for people of different leg length. This study proves, Gibsonian ecological psychologists claim, that the affordance "climbability" is perceived (Blau & Wagman, 2023, chapter 5). Moreover, they claim that the perception is direct, because the pattern specific for the affordance is available in the physical relation between the length of the step and the leg. Perceivers just need to be sensitive to that pattern to perceive it directly.

Gibson stresses that even babies are responsive to meaningful patterns but acknowledges that sensitivity to affordances develops and keeps on evolving. Organisms refine how and what they perceive throughout their lives. They learn to notice invariants that they did not notice before. In collaborative work with his partner, Eleanore Gibson (1910–2002), J. J. Gibson elaborated on this developmental process of perceptual refinement, and in Gibson (1966) the term "the education of attention" was coined. So, from its start, the ecological approach to perception has seen an important role for learning and development in becoming sensitive to affordances (Gibson & Gibson, 1955; Gibson & Pick, 2000). Ecological theorists don't construe learning as the acquisition of representational knowledge structures, but rather as progressive adaptation or attunement to the environment through reorganization of the organism-environment structure.

As noted, Gibson's ideas about perception have a Reidian flavor. Both see perception as of the environment directly, without needing mediating ideas, representations, or percepts. Both deny that sensations are the basis and cause of perception (Ben-Zeev, 1986). Gibson's notion of information in the ambient energy is similar to Reid's notion of visible figure (chapter 3, section 3.4.4). They both stress acquired sensitivity to patterns in changing stimulation as key to understanding how moving perceivers can perceive the environment. In fact, both assert that one sees the size of a table because of the ways what is available to the eyes changes when one moves with respect to the table. While a philosopher like Hume has invoked such changes as a reason for a mediated view of perception, both Reid and Gibson draw exactly the opposite conclusion. For them the availability of external or interactive patterns obviates the need for internal mediators.

Gibson offers a comprehensive theoretical framework for studying perception and has launched a program for empirical research with concrete results. Gibsonian ecological psychology very explicitly presents itself as an alternative to broadly cognitivist approaches to perception, cognition, and mind (Blau & Wagman, 2023). Ecological theorists typically reject all three pillars sustaining standard cognitivist thinking about mind and cognition: mental representation, mental computation, and internalism.

10.3.3 Dynamical Systems Approaches to Cognition

Besides J. J. Gibson's ecological work, developments in the understanding of complex dynamical systems and self-organization have also influenced scientific approaches to the mind that are critical of cognitivism. As we'll see at the end of this chapter, there has been convergence between thinking about minds along both the ecological and the dynamical lines. But let's first expand on dynamical systems approaches to cognition and mind.

According to the dynamical systems approach to cognition and mind, cognition and minds should be understood scientifically in terms of concepts belonging to the theory of complex dynamical systems. This will allow to conceive of mind and cognition as phenomena of self-organization. Many promoting the dynamical systems approach to cognition hold that this way of understanding cognition and mind is incompatible with the core ideas of cognitivism.

To begin with, what is a dynamical system? The notion is very broad and general as it concerns any system that changes in time. A ball rolling off a hill, a galloping horse, or a painter adding a stroke to their canvas are all examples of dynamical systems. The boundaries of dynamical systems are flexible and often dependent on the goals of the theorist. The theorist can consider the ball and the hill as a system, but the ball itself can also be seen as a system consisting of molecules. Again, the molecules can be considered as systems of atoms and so on.

In physics, a prominent way to study dynamical systems is by writing equations that describe how the system will change over time. Such an equation could, for example, specify precisely how a phenomenon of interest, such as the pressure a fluid exerts on the walls of a container, changes when factors such as the temperature of the fluid change. So far, so good. Dynamical systems can become *complex* when they have additional features, often occurring in combination. One prominent feature is nonlinearity of effects. Processes with linear effects are those in which changes in effects are proportional to changes in causes.

Imagine you're sitting in a canoe, gently rocking it from side to side. At first, the canoe responds calmly, returning to its original position each time. If you increase the intensity slightly, it still reacts in kind—adjusting to the motion and stabilizing again. But if you continue to rock it with growing force, there comes a tipping point. Beyond that threshold, the canoe no longer stabilizes—it capsizes. This illustrates a nonlinear shift in behavior: Up to a certain point, small changes produce proportionate effects. But once that threshold is crossed, even a minor increase can trigger a dramatically different outcome.

Though the rocking canoe serves well to illustrate nonlinearity of effects, it does not concern a phenomenon that in itself shows much of the interesting complexity of cognition. But another phenomenon that on first sight might look mundane has what it takes: finger wagging. Readers can try for themselves. Just put your hands in front of each other with your fingers pointing to each other. Now start wagging your left hand slowly up and down. Next, do the same with your right hand, with the same speed. If you experiment a little, you'll notice that you can start the right-hand wagging whenever you want. The left and right hand cycles have no other relation than having the same length. When your left hand is at a certain point of its cycle, your right hand could be anywhere in its cycle, depending on when you choose to begin your right-hand wagging. Now turn up the speed. You'll notice that, as you do this, the ways the cycles of the two hands are related will become constricted to two options. Either you'll be wagging out of phase, meaning that when one hand is up, the other is down, or you will be wagging in phase, both hands up and down together. If you are wagging out of phase, and you still turn up the speed, there will remain only one option: in-phase wagging. This is the famous finger-wagging phenomenon modeled dynamically first by Haken et al. (1985). It has become a simple but telling example of a complex dynamic system, because some of the core properties of such systems are present in it.

Finger wagging shows nonlinearity of effects. At some point a small increase in speed leads to a big change in effect: the reduction of cycle relations to only two or one options. Finger wagging also illustrates another aspect of complex systems relevant to cognition and mind: coupling. Coupling in this context refers to a fundamental kind of dependence of one factor on one or more other factors. In the wagging example, at high speed, the hands are coupled. Mathematically, coupling is visible in the equations that describe the system. The value of variable A will depend on variable B, while the value of variable B will depend on the value of variable A. But there is more. Notice that the finger-wagging phenomenon has been described in terms of three conditions that characterize the system consisting of the two hands as a whole. The cycles of the hand can be in phase, in antiphase, or, at slower speed, in a different phase relation. In this way, the finger-wagging example illustrates how a simple characterization of the system as a whole, or at the collective level is possible by means of applying the labels "in phase," "in antiphase" or "out of phase." These labels are called the values of an "order parameter," where an order parameter is understood as a general collective condition of the system at a certain moment or period in time. The values of the order parameter change as the speed of the wagging changes. Speed is a control parameter: something that determines which value the order parameter has. Further, finger wagging exemplifies the role of *attractors* in dynamical systems. Attractors are global configurations—particular values of an order parameter—of the system that the system will tend to evolve to. An attractor is a condition that the system will settle in under certain circumstances. Systems have multiple attractors, sometimes present at the same time. Attractors can change depending on circumstances. For example, in the finger-wagging example, when the value of the control parameter for speed grows, the system will tend to one of the "antiphase" and "in-phase" attractors and eventually to the "in-phase" attractor. When a system has evolved to one of its attractor conditions, it will tend to stay in that condition. It will require some effort to lead the system away from its attractors. In dynamical systems terms, the system needs to be "perturbed." One can think of attractors as valleys or pits in a hilly landscape. Things on the hill will tend to end up in the valley. It takes effort to move things out of pits. The deeper the pit, the more effort is required. Analogously, attractors can be stronger or weaker, and stronger attractors require a stronger perturbation.

The finger-wagging example also allows to illustrate a property crucial to many dynamical system approaches to cognition: It displays *self-organization*. That is, the value of the order parameter eventually obtained is dependent upon the speed of the wagging, but it is not *prescribed* by the speed of the wagging. At a certain high speed, both hands will move in phase. But nothing in the speed value at that critical point prescribes or instructs the hands to start moving in the in-phase way. The speed doesn't tell or command the hands how to move. How could it? So, though the particular organization obtained by the hand is orderly—namely, in phase—that order does not derive from a cause that describes that order. Neither does the speed, or its critical value have that order itself, so that it could be "imprinted" upon the hands.

In other words, the term "self-organization" is used because the orderly pattern or organization displayed by the relative movements of the hands is not the result of an external preexisting source that prescribes that organization. A system self-organizes if it forms patterns that are novel in the sense that they are not prefigured or dictated by their causal conditions. In a sense, the elements of the system "fall into place," "settle," or "find a balance."

That dynamical systems can show self-organization plays a key role in seeing dynamical systems approaches to cognition as opposed to cognitivism. Indeed, self-organization had played the same role in earlier debates in embryology (surveyed in Thelen & Smith, 2006; see also Withagen, 2022, chapter 6). Embryology concerns how a single fertilized egg turns into an embryo with all its intricate structures. The dynamical way of thinking in embryology arose as a reaction against *instructivism*. Instructivists, encouraged by the progress made in the science of genetics in the twentieth century, thought of embryogenesis as rigidly controlled by genes. Genes are then seen as containing a "blueprint" for the developing organism, which acts as a program that steers the growth of the embryo along a predetermined path that the genes "code for." Ever since the second half of the twentieth century, biologists have criticized this way of thinking about development and have argued against the metaphors of "code" and "instruction." We have already met, in section 4.7 of chapter 4, key figures in this line of anti-instructivist thinking, such as Conrad Waddington and Susan Oyama (Oyama, 1985; Waddington, 1957).

They have defended views of embryogenesis as basically a process of dynamical self-organization. On such a view, genes are one, very important, factor in the process of ontogenesis, but far from the only one. That is, genes don't instruct, but trigger or influence processes that happen only because many other factors also contribute. The dynamical self-organizational view in embryology hence endorses multicausality: Control over development is seen as spread in space and time. For example, at the beginning, very similar cells migrate to different loci. Because of their different surrounding, they develop differently. Once this initial differentiation has occurred, it influences further development. The place where the initially similar cells have arrived develops along a specific path. And the acquired specificity at that locus influences further processes in the developmental cascade. Here, we have an example of what in dynamical thinking is called nestedness of spatiotemporal scales. What happens on a local spatial and temporal scale (here and now) is determined by the conditions that have been laid down at more encompassing temporal and spatial scales (Thelen & Smith, 2006).

To summarize, in the dynamical view of embryogenetic development, along the way local and global patterns emerge that then play a role in the further direction of development. Some further processes are enabled; other processes are rendered impossible by conditions that have been established during the process. On this dynamical view of embryogenesis, there is never a central controller but an unfolding in which every step is made possible by the previous steps. Genes retain a causal role in embryogenesis, but not an exclusive one. Hence, the whole process is determined by a multiplicity of factors that are spread and nested in space and time. Multicausality rules.

Some dynamical systems theorists have argued that lessons from embryogenesis apply to action and cognition, too (Thelen & Smith 2006; Withagen, 2022). The resulting dynamical systems approach to action and cognition sees action and cognition as permanently developing multicausal processes of self-organization. In the dynamical view, action does not happen because actors follow a description that prefigures or a program that guides the performed action. Similarly, cognitive activity is intelligent not because it is steered—instructed—by some kind of descriptive knowledge stored within the system. Rather, agents act intelligently because they have developed ways of acting and interacting throughout their history of being coupled to an environment. During this history of coupling, structure emerges at a number of spatiotemporal scales. Larger-scale processes determine what happens on more narrow scales. Sometimes this determination occurring at a larger scale is conceived of as a *constraint* (Juarrero, 1999, 2023). Because of conditions that have emerged at a more encompassing spatial or temporal scale, development is pushed in a certain direction. One should not think of constraints as limitations. Constraints create possibilities: further conditions that would be impossible without the constraints. A concerto pianist can achieve their mastery only if they have been through a regime of training that has been setting the conditions, over time, for their current virtuoso performance. The training occurs on a broad temporal scale and acts as a constraint on how they are playing now.

In the dynamical view of cognition, intelligent agents come to embody knowledge. They manifest knowledge, without relying on separate "knowledge systems" within them that store knowledge and impose it on "mere movement" systems. Instead, on the dynamical view, cognition is a permanent process of self-organization. Stable patterns occur—attractors—but these are always the result of a multiplicity of factors, and always open to further change. Attractors can shift or, under different conditions, break down, and make room for new attractors (Blau & Wagman, 2023, pp. 238–242).

To make this more concrete, it is helpful to look first at an instance of how the dynamical systems view has been applied to the development of human activity. One developmental behavioral phenomenon is how small infants become upright walkers. This process has been studied through a dynamical systems lens in great detail by Esther Thelen, Linda Smith, and collaborators. They have found that the development of upright walking shows all the signs of a dynamical system (Thelen & Smith, 1994). That a child acquires the capacity to walk upright is the result of multiple factors, both inside the child's changing body and its environment. Walking occurs when all the contributing causes act in concert to allow for it. The required balance can be temporarily obtained in a fragile form and disappear again when one of the contributing factors changes during development. For example, consider the puzzle of the apparent disappearance of the stepping response. Newborns, when held upright, will often make stepping movements with their legs. However, they cease to do so after a few months. Attempts at explaining this have assumed, for example, a suppression mechanism in the brain. But Thelen and Fisher (1982) provided a more parsimonious dynamical explanation. In their first months, children gain body fat. As a result, their legs gain weight and can no longer be moved by the muscles that were able to initiate the stepping pattern before. This explanation was confirmed when it was observed that the stepping reappeared when the infants' legs were submerged in water. Under these circumstances, the muscles can get the job done with less force (Thelen et al., 1984).

Overall, organization results from a multitude of factors exerting influence. These factors change over time and hence the overall organization changes. Some forms of organization can become very stable. They are strong attractors. But even the most stable configurations in dynamical systems remain susceptible to change—otherwise, the system would no longer be a dynamical one. Perhaps a lot has to happen before a stable pattern changes, but change is always possible. In fact, in living systems, change is inevitable. Therefore, the kinds of self-organization seen in the development of living systems is sometimes described as "soft assembly." The patterns that are formed can be stable, but they are never completely fixed (Thelen & Smith, 2006, p. 274).

Esther Thelen and Linda Smith have also been pioneers in establishing the dynamical approach to cognition, over and above action, as an empirical discipline (Thelen & Smith, 1994). One of the phenomena they have investigated from a dynamical perspective concerns the so-called "A-not-B error." This term names a remarkable pattern observed and theorized initially by Swiss developmental psychologist Jean Piaget (1896–1980). A typical experimental task involving the A-not-B error will run as follows. A child is placed in front of two boxes with identical lids, at locations A and B. First, an experimenter shows the child a toy. Then, visible to the child, the experimenter hides it at location A, and the infant is allowed to reach for it. After a number of such A sessions, the experimenter hides the toy at B, then again allowing the infant to reach. If there is a short delay between the hiding at B and the reaching, eight- to-ten-month-old infants often commit the A-not-B Error: They reach for A, even after the toy was visibly hidden at B. The error disappears gradually from the behavior of older children. Infants older than ten months on will tend to reach correctly at B in the B condition, that is, the place where they saw the toy disappear.

Interestingly, several contextual factors are known to influence the occurrence of the A-not-B error. Thelen et al. (2001) mention: "The visual properties of the hiding locations, including the distinctiveness, distance, number, and transparency of the covers . . . whether search involves reaching or just looking . . . whether there are landmarks in the environment . . . whether infants search for objects, food treats, or people . . . whether the task is done at home or in the laboratory . . . whether the infant or the hiding places have been moved . . . and the infants' amount of crawling experience (Thelen et al., 2001, p. 3).

Explanations of the phenomenon in line with traditional cognitivism might appeal to the hypothesis that the crucial difference between infants younger or older than ten months concerns a specific cognitive factor. So, one can assume that object knowledge has not yet reached its mature level, that spatial cognition is underdeveloped, or one might try to explain the data by involving the failure, in children less than twelve months, of cognition to inhibit incipient action. In all these explanations, cognition is separated from action, and an explanation is sought in the relation between cognition and action. Cognition itself is not yet developed, or it is, but it fails to reach action. In both these options, cognition and action are viewed as two separate things that can or fail to causally interact.

Such explanations, Thelen and her colleagues argue, run into problems because of the context effects (Thelen et al., 2001). Why does cognition suddenly appear in younger children when the delay is shortened, when the task is performed with more distinctive hiding places, or when the child is upright instead of sitting in the B condition, for example? Instead of invoking cognition as a separate cause, distinct from action, Thelen et al. (2001) explain the observed patterns in behavior in terms of an interplay between such factors as the timing and the length of the delay; where the child has previously looked at; the visual features of what they looked at; where the child has reached before; and how many times they have done so. These factors have a relative strength, and they influence each other. The transition occurring from ten months on is explained as a change in relative weights and mutual influence between such factors. These changes in weight are not left unexplained. For example, the fact that infants increasingly self-locomote from the end of their first year on, is invoked as a possible explanation for why looking at a different spatial location acquires a stronger impact on spatial behavior, including reaching (Thelen et al., 2001, p. 10).

The model allows to include unexpected bodily specifics. For example, young children perform above what's expected from them when they are held upright in the B condition. They also show correct performance when they are reaching with a weight on their wrists in the A condition, but not in the B condition (Thelen & Smith, 2006). In the model, it is assumed that such factors change the effect of past A condition behavior on current B behavior. Thelen and her colleagues assume that changing bodily conditions in the B condition changes the effects of past reaching history. Interestingly, it can be noted that this assumption resonates with the finding reported earlier in this chapter, that remembering tends to be facilitated by taking a body position congruent with what happened in the remembered event.

Reminiscent of Dewey's warnings not to commit the historical fallacy (see chapter 3, section 3.4.5), Thelen and Smith see the knowledge that is acquired by a child as a product of development, rather than as constant cause of it.

Moreover, knowledge and action are not distinct. The development of action and of knowledge are sides of the same coin. For example, children are not assumed to have already complete knowledge that their motor systems fail to access, but in a process of gradual development they become knowledgeable agents. Hence the dynamical systems treatment of the A-not-B error by Thelen and colleagues endorses the view that cognition is fundamentally embodied and embedded. The relevant knowledge children arrive at is acquired through and manifested in knowledgeable action-in-a-context.

The work by Thelen, Smith, and colleagues on the A-not-B error has become a celebrated example of the dynamical systems approach to cognition. But it is just one example of what is now a broader research program, tackling many phenomena from problem solving to language (see, e.g., Dixon et al., 2014, Van Orden et al., 2001). Many, though not all, theorists that espouse a dynamical systems approach to cognition see it as an approach that is incompatible with cognitivism. They harbor a view of action, cognition, and intelligence as *in* interactions of embodied agents rather than due to mental cognitive operations or relying on stored mental representations. Agents can interact intelligently because their brains and bodies have come to be organized throughout a history of interactions with an environment it is coupled to. On this dynamical view, explanations in terms of representations and cognitive operations should be replaced by multicausal analyses that involve nonlinear self-organization occurring at multiple temporal and spatial scales. Forms of organization obtained at broader temporal and spatial scales constrain forms of organization that can be obtained on finer temporal and spatial scales (Juarrero, 1999, 2023). Organization thus encompasses history, the environment, and the body. Cognition becomes distributed in time and space. Internalism and neurocentrism—the idea that the mind is in the brain—are rejected. The work that in traditional approaches would be done by cognitive operations in the head is reconceived in terms of dynamical changes in a self-organizing system comprised of an organism and the environment it is coupled to. In these interactions, agents, and environments are as tightly coupled as the hands are in the finger-wagging example. Moreover, the brain is only one factor in these interactions. It never operates on its own in cognition, but only in concert with organic processes, bodily structures, and activities, all taking place in an environment with a rich spatial and temporal structure.

Dynamical system theorists that oppose cognitivism are labeled as advocates of "radical embodied cognitive science" by Tony Chemero (2009). This radical embodied dynamical systems approach to cognition shares its rejection of the main tenets of cognitivism and its construal of cognition as world-involving activity with the Gibsonian ecological approach. It should come as no surprise then, to hear that many dynamical systems methods are perfectly suited to pursue the Gibsonian scientific program (Blau & Wagman, 2023, part 3). A resulting dynamical ecological view of the mind has emerged that relies on both dynamical and Gibsonian concepts. It is not restricted to perception but investigates cognitive phenomena such as thinking, learning, and remembering (Blau & Wagman, 2023, chapter 12).

These developments are natural. The ecological and dynamical theorists discussed in this chapter endorse a similar picture of the mind and cognition. They see mind and cognition as products of interaction, permanently open to further change and development. They endorse multicausality, and the involvement of a multiplicity of factors, extending in space and time. Hence, the body, environment—including the social sphere—and history become partners in making the mind into what it is.

To see an example of this ecological/dynamical conception of minds at work, consider how Käufer and Chemero describe how people act in environments when engaged in specific skilled activity. When engaged in a task, we tend to attend only to some aspects of the environment. Or, in Gibsonian language: Only some of the many possible affordances appeal. Here is how Käufer and Chemero characterize the situation:

> When a human or other animal is engaged in a task, it organizes itself into a temporary, special-purpose dynamical system—a bike rider, a finger wagger, a roadkill scavenger—and as such it is only sensitive to the affordances that are relevant to that temporary, special-purpose dynamical system. Only those relevant affordances that can perturb the temporary, special-purpose dynamical system are experienced as invitations. (Käufer & Chemero, 2021, p. 245)

It is important to note that self-organization is tied to certain physical conditions. Self-organization is typically found in physical systems that are far from thermodynamic equilibrium (von Förster, 1960). Thermodynamic equilibrium refers to equal distribution of energy over different components. In total equilibrium, every part of a system has the same amount of energy. As a result of that, nothing happens. For something to change, there must be some flow of energy between areas of different energy level. But if all energy is equally spread, there are no such areas. Equal distribution of energy is also called "maximal entropy." Maximal entropy is the opposite of order, exactly because when entropy is at its maximum, nothing happens. According to the so-called second law of thermodynamics—formulated, by the way, by Hermann von Helmholtz (see chapter 3)—the total entropy of the universe is always increasing. This means that, over eons of time, the universe will end in maximal entropy.

Self-organization occurs under away-from-equilibrium conditions. In general, degree of organization is inversely related to degree of equilibrium. Living systems are always in a far-from-equilibrium condition. As said, getting closer to equilibrium is a loss of order, and living systems need order—in the form of differentiated organs for example—to stay alive. To remain alive, living systems need to remain in that far-from-equilibrium condition. However, according to the second law of thermodynamics, the overall entropy of the universe must increase. Therefore, living systems must increase the entropy, or decrease the order, in their environments. In this way, the overall entropy of organisms and environments together increases, and the second law of thermodynamics applies globally. So, via thermodynamics, self-organization can be linked to far-from-equilibrium systems, and from there to living systems.

Bibliographical Note for Chapter 10

The literature on the extended mind is extensive and has found its way to other introductions to the *Philosophy of Mind*, see Bayne (2022), chapter 7. The idea that artifacts transform the mind, present already in Clark and Chalmers (1998), has been pursued at book length in Menary (2007).

Two historically important strands in "radical" views of embodiment and dynamics I have left out are the work in behavior-based robotics often associated with Rodney Brooks, and the dynamical system-based criticism of cognitivism of Tim Van Gelder. Their landmark papers remain highly readable (Brooks, 1991; Van Gelder, 1995).

A shorter introduction to the Gibsonian approach, compared to the comprehensive Blau and Wagman (2023), can be found in Segundo-Ortin and Raja (2024). For an overview of different strands in Gibsonian thinking, see Bruineberg et al. (2023). The Gibsonian approach has been combined with Wittgensteinian ideas in influential publications by Erik Rietveld and colleagues (see, e.g., Rietveld, 2008, and Rietveld & Kiverstein, 2014).

J. Scott Kelso's *Dynamic Patterns: The Self-Organization of Brain and Behavior* remains an excellent introduction to the field of dynamical systems approaches to cognition (Kelso, 1995).

Last but not least, I want to mention a book that inspired the title for this chapter: Louise Barrett's *Beyond the Brain*, offering an embodied and embedded view of cognition by a researcher with deep knowledge of animal cognition (Barrett, 2011).

11 Life and Mind

> But the mind is, from its very nature, a living and active being. Every thing we know of it implies life and active energy.
>
> —Thomas Reid (1785/2002, p. 21)

11.1 Enter Life

Many of those who have thought about the mind have not focused on organic life. This is remarkable, because as far as our experience goes, to have a mind and to be alive seem to be connected in fundamental ways.

Aristotle recognized such a link between mind and life by ascribing a vegetative soul to all living beings, including plants, and a sensitive soul to all animals (see chapter 2). Yet, when it came to the pinnacle of mentality, the possession of a rational mind, life ceased to be of much theoretical relevance. Descartes's res cogitans was fundamentally nonorganic, despite it being able to enter in a tight relationship with the organic—as when one's pure thought of the mathematical essence of a circle is coupled to a mental image of one (chapter 2).

It is not necessary to commit to a stark or explicit dualism to sideline life. It can happen by objectifying mental phenomena and abstracting from the conditions under which these phenomena naturally occur, for example, when perceiving is conceived of in terms of impressions, ideas, sensations, percepts, or perceptual states.

Philosophers such as Dewey and Wittgenstein warned for excessive abstraction and objectification. In their view, understanding a mental phenomenon is not achieved by isolating it from its natural context but, on the contrary, by examining how it comes about within the totality of that context. From this perspective, the puzzles faced by thinking about consciousness in terms of qualia or about meaning in terms of a picturing relation become predictably unsolvable. If one thinks of the quale of red as an intrinsic, nonfunctional property, one can never reinsert it into the natural world. Then, one faces a hard problem of consciousness indeed. Similarly, if one thinks of the meaning relation from words to world without considering the use of words, one is left without any means to answer the question how the intentional relation between language and reality is possible.

Wittgenstein and like-minded thinkers pointed to the relevant circumstances with the concepts of "life forms" and "natural history," and insisted that one could not understand minds without understanding life, as illustrated by a comment on a remark by Anscombe:

> "I have listed 'men' with such objects as 'cats' and 'turnips,'" Elizabeth Anscombe wrote in 1944, insisting that any attempt to understand ourselves must begin from the fact that we are living creatures. (Mac Cumhaill & Wiseman, 2023, p. xiii)

Yet, neither Anscombe nor Wittgenstein *theorized* life or its connection to mentality. Dewey did, showing himself to be a Darwinian, evolutionary philosopher (chapters 3 and 4, see also Popp, 2007). Darwin himself engaged in linking evolution, mind, and life. Besides writing on the expression of emotion (Darwin, 1872), he investigated and was surprised by the intelligence shown by the lowly earthworm (see chapter 4). In science, evolution was invoked to explain mental phenomena by Skinner's selection-by-consequences approach (see chapter 4). In philosophy, evolution and selection are crucial in Ruth Millikan's teleosemantic attempt to explain intentionality and representation (see chapter 9).

Yet, all these approaches focus on evolution, rather than on life. In the 1990s "enactivist" philosophers developed the idea that both consciousness and cognition could be understood only by realizing that it shares fundamental features with the phenomenon of life. They thought cognitivist science and philosophy had made the mistake of divorcing consciousness and cognition from life. The enactivists endorse a "life–mind continuity thesis," formulated and defended in Thompson (2007, building on Varela et al., 1991). The original enactivists start out from a philosophical analysis of the phenomenon of life and claim that the structure of life reappears, be it in modified ways, in crucial aspects of mind. That is, there is a structure that is common to single-celled organisms, the sensorimotor coupling of organisms and environments, social interactions, and language. All of these phenomena are characterized by what enactivists call "autonomy" (Di Paolo, 2005; Di Paolo & Thompson, 2014). Let us follow this growth and diversification of autonomy from cell to language to see how enactivists like Ezequiel Di Paolo, Hanne De Jaegher and Elena Cuffari have elaborated the life–mind continuity thesis (Di Paolo et al., 2018).

11.2 Spiraling Autonomy

11.2.1 Basic Autonomy

Thomas Reid believed in a fundamental difference between the living and the nonliving. Life was active, matter passive. Living things move by themselves. Natural objects move because they are causally impacted; artifacts move because they have been set up by living things to move. Reid understood this difference in dualistic, theological terms. God created life in addition to matter and gave it special nonmaterial properties—"active energy" (Reid, 1785/2002).

Enactivists too endorse a fundamental distinction between what lives and what doesn't. Yet, they don't reject a broadly materialist, naturalistic view. Living systems, in their view, are physical systems. However, they are a special kind of physical system, with properties that other kinds of physical system lack. In particular, living systems have characteristic forms of organization and operation.

Enactivists have designated what they consider to be the defining properties of life by the concepts of *autopoiesis*, *operational closure*, *precariousness*, and *autonomy*. To explicate these concepts, let's consider the simplest lifeform: a single cell. A cell consists of a cell wall, or membrane, that contains liquid with molecules floating in it. Some of these molecules are very complicated and perform specific jobs. They can be seen as similar, though on a much smaller scale, to bodily organs, like liver, heart, and lungs. Together, the constituents of a cell are continually engaged in processes of self-production or *autopoiesis*, say the original enactivists. Autopoietic processes are characterized by a specific tight form of interdependence between the activities that together make the process as a whole happen. Think about the cell membrane and the processes occurring inside it. If there was no cell wall, the molecules inside the cell would be flooded, and washed away and broken like the furniture of a house in a tsunami. However, the cell wall is not a static structure like the wall of a house or a dike. A cell membrane, like your skin, is itself a complex structure that needs to be continually maintained, repaired, and rebuilt.

Many processes necessary for the cell membrane's continued existence occur on the inside of the cell but not in the membrane. But we started by noticing that these intracellular processes are possible only because of the cell membrane's existence. Hence a circular organization ensues.

A bit more formally, in the autopoietic processes, the activity of constituent A is necessary to produce constituent B, but the activity of constituent B is necessary to produce constituent A. This implies a form of circular interdependence characteristic for autopoietic processes. This intense form of interrelatedness goes beyond two constituents and might involve arbitrarily many more steps. A might be necessary for B, B necessary for C, and C necessary

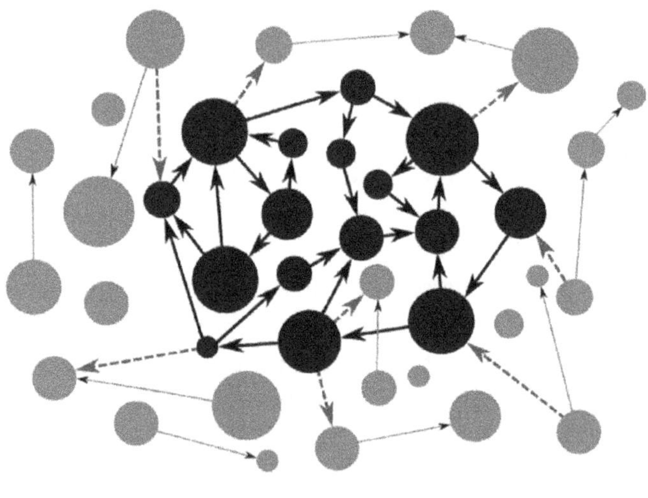

Figure 11.1
Operational closure. Processes are drawn as circles. The black ones form an operationally closed network because they are both producing and produced by other black circles. They have both incoming and outgoing arrows to and from other black circles. Gray circles are connected to black circles only by arrows in one direction. Source: Di Paolo et al. (2018, p. 26). Copyright Ezequiel Di Paolo (2013). This work is licensed under a Creative Commons Attribution-NonCommercial-ShareAlike 3.0 Unported License. http://creativecommons.org/licenses/by-nc-sa/3.0/deed.en_US

for A, and many more intermediate steps might be introduced. The original enactivists used the term "operational closure" to describe the autopoietic circularity typical of living organisms, from single cell to whole animal.

The specific relations between the constituents of an autopoietic, operationally closed living system make it into a unique individual, different from all other individuals. But operational closure does not imply that living organisms are completely self-standing. On the contrary, living systems cannot exist without an environment that they interact with.

Reconsider the cell membrane. Any cell membrane, be it of a single-celled organism or of a cell forming part of an animal's body, finds itself in an environment or milieu with which it intensely interacts. The cell membrane allows molecules from the extracellular environment to enter the cell. Going in the other direction, the membrane acts as a gate through which molecules, for example, waste or byproducts of vital reactions, can be released into the world outside the cell. However, the relations between the cell and its environment are different from the intracellular processes. The nutrients that flow into a cell from its environment are not dependent for their existence upon the cell's activities in the way that the cell's membrane is dependent upon the continuation of intracellular processes. There would be carbon dioxide in the environment even if all plants that use it for photosynthesis would cease to be. Original enactivists used the term "structurally coupled" to indicate the specific relation, different from operational closure, between organisms and their environments.

The ways that organisms are structurally coupled to their environments is determined by their need to self-generate. Autopoiesis happens only when a specific ensemble of operationally closed processes occurs, and this happens only in some environmental milieus.

The self-production of living systems is therefore *precarious*. Changes to either the internal or the interactive processes sustaining a living organism can easily lead to the cessation of the whole process. Take out the energy transporters inside the cell and every other process will grind to a halt. Move a breathing organism to an oxygen-free environment and it dies. Living systems are always on the brink of degrading and, ultimately, dying, and have to continually work to stay afloat. Enactivists call systems that show both operational closure and precariousness *autonomous* systems (Di Paolo et al., 2018, p. 25).

Operational closure, so enactivists claim, means that there is a nonarbitrary distinction between the system and its surroundings. Everything that belongs to the operationally closed network of processes belongs to the individual. By continuously rebuilding itself, the system self-individuates: By its own activity, it becomes distinct from its surroundings. A distinct precarious system also implies a perspective from which norms arise. Being a living system implies a differentiation between what supports, undermines, or remains neutral with respect to its continued existence. It becomes possible to evaluate what the system relates to as good, bad, or neutral.

Enactivists hold that such norms and having a perspective do apply to autonomous systems by virtue of their being both operationally closed and precarious. Other physical entities or systems, such as stones or machines, don't have the required structure. They are neither operationally closed nor is their existence precarious in the required sense. It makes no sense to ascribe a perspective to them, or describe things as being good or bad for them. Enactivists further hold that only some kinds of physical systems can have the kind of organization necessary for autonomy and precariousness. If autonomy and precariousness are necessary for life, and if life is necessary for mind, mind is no longer substrate-independent in the sense in which it was for defenders of wide-ranging multiple realizability (see chapter 8).

Enactivists claim that the autonomy—in the technical sense of the word—of living systems gives them a basic form of subjectivity. By being autonomous, living systems inhabit a world that is significant to them, in which they care, and show concern for things. Di Paolo (2005) and Di Paolo et al. (2018) describe a crucial role of agency in the emergence of this form of basic subjectivity and sense-making. Agency as they define it involves the adaptive adjustment of an organism's interactions with an environment in light of how the circumstances are "good and better, bad and worse" (Di Paolo et al. 2018, p. 33). To be able to do so, organisms must be sensitive to how their current interactions affect and will tend to further affect their own viability and integrity, and redirect these interactions accordingly. Living systems adaptively modify their coupling to their environments, and thereby show basic sense-making and subjectivity (Di Paolo et al., 2018, section 2.4.3).

11.2.2 Sensorimotor Autonomy

Living creatures interact with their environments. This can occur in very basic ways, as in single-celled organisms. But as the tree of life kept growing, organisms became sensorimotor agents. Some organisms acquired sensory organs that allow them to perceive their environments, and tentacles, limbs, and other organs to allow complicated activity in their environments. Perception and action are tightly connected. A sensorimotor organization ensues.

Enactivists have claimed that this sensorimotor organization acquires the kind of autonomy they have identified in the organization of the living body (Di Paolo, 2005). The sensorimotor organization is autonomous because it is operationally closed and precarious. To characterize this autonomous sensorimotor organization, Di Paolo et al. (2017) take their lead from the so-called "sensorimotor contingency approach" to perceptual consciousness that has become mainly known through the work of Kevin O'Regan and Alva Noë (O'Regan & Noë, 2001). This sensorimotor contingency approach was developed in the context of the discussion about the explanation of phenomenal consciousness. O'Regan and Noë (2001) opposed the idea that the task of explaining consciousness formed an unsolvable "hard problem." Focusing on visual perception, they argued that qualia-based pleas for the existence of a hard problem of consciousness are built on mistaken views about visual perception. Hard problems about visual consciousness raise their heads when one thinks of vision in terms of the production in an organism's brain of an objectlike percept. When thinking of the percept as a pattern of neural activity, one will be led to ask how it is possible that this pattern has, besides its neural properties, also conscious properties. How does the neural firing generate the qualia?

These questions have no answers, according to the sensorimotor theorists, because they are built on wrong assumptions about visual perception and visual consciousness. Visual perception is not aimed at producing a percept inside a person. Instead of thinking about visual perception as "something that happens inside of us," say sensorimotor contingency theorists, it should be recognized as "something we do" (O'Regan & Noë 2001). Visual perception is like feeling the softness of a sponge by touch. One is able to experience its softness only by pressing the sponge and finding out that it offers little resistance. Further, visual consciousness is not something over and above what organisms do when they perceive—it is *in* the doings. It is not generated but "acted out." Again, the sponge can serve as an analogy. Feeling the softness is in the pattern of squishing; it is not a further product separate from it (O'Regan, 2011).

The sensorimotor contingency theorists have offered analyses of several visual phenomena to substantiate these claims. These analyses revolve around the concept of sensorimotor contingencies. These are ways in which sensory stimulation changes for an organism in relation to the movements that an organism makes—contingent doesn't mean accidental here, but rather dependent upon. Consider a black line against a white background. If a perceiver moves their eye along the line, as long as the line continues, the stimulation reaching the eye will not change. They can tell by patterns of sensorimotor contingencies that a line is seen. Sensorimotor theorists think in similar ways about the sensorimotor contingencies of visual properties, such as colors, or seen objects. To see red is to find out that certain patterns of change typical for red occur. To see a cube is to be engaged with an object in a visual interaction during which the pattern of sensorimotor contingencies specific for cubes occur. Moving around a cube creates a specific pattern: seeing the facing side first, then an angle between two sides, then one side again, an angle between that side and the back side, and so on. To engage in a visuomotor interaction with it in which such pattern unfolds, implies seeing a cube. The sensorimotor approach explains the feeling of visual completeness and detail: the feeling we have of seeing the whole scene we are looking at. When you look out in front of you, it feels to you as if you are visually taking in a whole scene. The traditional explanation for this feeling of completeness and detail is that a complete and detailed percept representing the whole scene in detail is built up in the brain. Sensorimotor theorists reject this explanation. Why does a complete percept give rise to feeling of completeness? Perhaps it *just does*, but then why can't one equally say that a complete outside scene *just gives rise* to a feeling of completeness when seen? If it suffices that something is complete, it can be the outside world as well as an internal percept. So, some extra story needs to be told about how a complete percept gives rise to a feeling of completeness. But then the internal percept option offers no advantage to the sensorimotor approach. Both need to account for how a feeling of completeness arises.

To see how the sensorimotor account of completeness goes, let's consider empirical material that sensorimotor theorists have invoked to support their theory. We often don't notice large changes in scenes in front of us, when our attention is distracted. This has been proved abundantly in "attentional blindness" experiments. In one widely known version, people watching a video of a group of basketball players don't spot a person in a gorilla suit entering the scene and stopping at the center, because they are busy counting how many times the ball gets passed (Simons & Chabris, 1999). Sensorimotor theorists have pointed to such experiments as showing that there is no internal percept formed. They have also cited attentional blindness as a phenomenon that provides evidence for a sensorimotor account of the feeling of completeness. Sensorimotor theorists explain the feeling of completeness by pointing to the interactive structure of seeing. Visual perceivers have the feeling of seeing a whole scene because they normally will attend to significant changes. If something suddenly appears in a scene, viewers will immediately focus their eyes on where the change happens. Not everything in a scene is simultaneously seen, but, standardly, any significant change is bound to draw attention. Therefore, visual perceivers have the feeling of having the whole scene in their "visual grip." Attentional blindness experiments exploit this interactive structure. They lock viewers' attention, for example, to the ball passing, so that attention can't do its normal change-tracking job. Therefore, the gorilla slips through.

The sensorimotor approach formed a starting point in a subsequent book coauthored by Ezequiel Di Paolo, Thomas Buhrmann, and Xavier Barandarian (Di Paolo et al., 2017). Drawing inspiration from earlier work by prominent Swiss developmental psychologist Jean Piaget, whom we already met in the previous chapter, they focused on patterns of sensorimotor contingencies, but extended their investigations to include patterns of patterns of sensorimotor contingency. For an example of such, consider a baby that learns to drink milk from a bottle. The baby needs to coordinate three actions, each coming with its specific sensorimotor profile. They must make the appropriate mouth movements so that milk comes out of the bottle, must swallow the milk, and must continue breathing (Di Paolo et al., 2018, p. 51). The overall coordination pattern must be coherent. The way of sucking constrains the way of drinking and the way of drinking constrains the way of breathing. There is an overall balance that must be achieved between parts of the overall pattern and finding that balance is solving a sensorimotor coordination problem. If patterns of sensorimotor patterns form coherent sensorimotor schemes, sensorimotor schemes cluster into yet larger patterns. Informally, individuals have styles of doing things that hang together and that mutually support and reinforce each other. Individuals have their "own ways" of eating, drinking, and gesturing in a conversation, and one can often recognize someone from seeing that person acting, even if one never saw this person performing that kind of action before. In the analysis provided by Di Paolo et al. (2018), the totality of interconnected sensorimotor patterns and schemes ultimately become autonomous, in the sense of the word introduced earlier. That is, they form a self-producing totality, that has the structure displayed in figure 11.1.

Informally, this autonomy shows in the fact that one element in a sensorimotor scheme tends to bring forth the next one, or that something done in one's particular style induces doing something else in the same style. Like maintaining autopoietic autonomy requires agency, or the active regulation of the interactions with the environment, so does maintaining sensorimotor autonomy. Sensorimotor organisms use their sensorimotor capacities to preserve the structure of their sensorimotor organization. The more skilled they are, the better they can do so when circumstances are different than usual or adverse. The better musician will be better able to improvise, to adapt to mistakes by themself or other players, to a different instrument, and so on.

The autonomy of sensorimotor bodies has important implications for how they relate to biological bodies. A sensorimotor body is said to be enabled and constrained but is not fully determined by the living body (Di Paolo et al., 2018, p. 48). Take the sensorimotor practice of smoking (Di Paolo et al., 2018, pp. 47–48). One can only smoke (in the appropriate sense) if one is a living body; smoking sustains further smoking, but it runs against the continued existence—the autonomy—of the living body. Despite being made possible by the self-producing activity of the living body, it has a tendency to annihilate that self-producing activity. Enactivists point out that the fact that their analyses of style and phenomena like smoking resonate with how people experience and describe their experience, validates their analyses. For example, when engaged in some activity we perform very well, done in one's own way, people have a feeling of agency (Di Paolo et al., 2018, section 3.1).

Enactivists stress that both the autonomy of living and of sensorimotor bodies are, unlike computational processes in the cognitivist picture, not independent of matter. Only some physical constellations can acquire the specific forms of organization leading to bodily (or somatic) and sensorimotor autonomy. Moreover, even if sensorimotor autonomy is not fully determined by somatic autonomy, sensorimotor autonomy is possible only when somatic autonomy is in place. Bodily autonomy is a necessary but not sufficient condition for sensorimotor autonomy. Enactivists point at dynamical systems (chapter 10) as having properties that parallel those crucial in their own analyses of the common structure of life and mind. Like living and therefore cognitive systems, dynamical systems show context-sensitive self-organization and flexible adaptive structures appear at different spatiotemporal scales simultaneously, in circumstances of operation removed far from thermodynamic equilibrium (Di Paolo et al., 2018, chapter 3).

11.2.3 Social Autonomy

A further form of autonomy discerned by enactivists, besides organic and sensorimotor autonomy, is social autonomy. Social autonomy emerges out of interactions between sensorimotor agents. The organic and sensorimotor organization of human beings is inclined to such participation. We already start participating with other people in perceiving them. We don't observe their movements neutrally, but as centers of actions that we ourselves can or could perform, or that could complement our own actions once we participate. Enactivists use the French phenomenological philosopher Merleau-Ponty's term "intercorporality" to designate this form of "bodily perception of another body" (Di Paolo et al., 2018, p. 63; MacLaren, 2002, p. 190).

Full-blooded social autonomy emerges interactively. When people—or possibly other organisms—meet, forms of interaction can happen that become intertwined or coordinated. An example given by enactivists is "the corridor dance" (Di Paolo et al., 2018, p. 69). This occurs when two people coming from opposing sides must pass through a relatively narrow passage. Often, they both simultaneously move to the same direction, only to block each other's way, to then repeat the same thing on the other side. Enactivists use the example to show that coordinative patterns often can be characterized as autonomous (Di Paolo et al., 2018, p. 65)—in a way comparable to how living and sensorimotor bodies are autonomous. Autonomy means that the coordinative patterns, even if they are enabled and constrained by the aims of the individuals, are not fully determined by these. In the corridor example, one can see how individual aims and collective outcome clash. They both want to pass, but because of the pattern they weave together, they can't.

Coordination can take a more complex shape when participants attempt to *regulate* the interactive pattern. In the corridor example, this would happen if one of them would initiate a sideways movement very slowly, so that it was out in the open for the other participant which direction they would take. By coordinating their mutual cooperative actions so that individuals can achieve things together that they could not achieve individually, coordination becomes what enactivists describe as "participatory sense-making." (Di Paolo et al., 2018, section 4.3).

To coregulate the coordination, participants can engage in what enactivists call "social acts." Social acts are built from partial acts: actions that are incomplete when performed by only one individual. Partial acts that have a characteristic complementary act—taking and giving, forwarding a hand and grasping it—succeed only if another participant partakes in it in a certain way. Giving something or initiating a handshake are examples of social acts. To succeed in giving something, or enact a handshake, someone else must take the gift, or accept the handshake. Social acts, thus, come with a shared know-how of how to enact them according to norms of interaction—norms that when met continue and enhance the interaction.

Like somatic and sensorimotor autonomy, social autonomy is analyzed by enactivists as precarious. It will break down if not actively supported by the participant's activities, and even if it is so supported, success is never guaranteed. Social acts offer means for containing the precariousness, but they can hamper coordination too. For example, the same overall partial act can diverge into different varieties. There are many ways to perform a greeting with the hands: shakes, fists, hands-up movements, and so on. This creates a problem of choice. Which one to initiate in which circumstances, and how to answer a given one? A means to constrain such situations of choice, and other potential forms of breakdown in cooperation, is to introduce coordination of the coordination via meta-regulatory acts. These are acts, or ways of acting, that steer which social acts are performed. For example, one might modulate the timing of stretching out one's hand or make a movement with one's other hand to draw the attention of another person to the kind of complementary act that they should engage in. Repertoires of social acts, including regulatory acts—both of which can be partial acts—build up in a community.

This enactivist treatment of social interaction, thus, describes a buildup of complexity in social interactions, with increasing regulation. The enactivists do warn about reading too much into the processes described. They claim that, in the most basic forms of social interaction—think about the corridor dance—only a very minimal sensitivity for self and others is present (Di Paolo et al., 2018, pp. 47–48, 162). Participants don't see, let alone conceive of, each other as a "self" standing over and against an "other." To overinterpret basic social interaction in this way goes against the methodological aim of making "minimal assumptions" the enactivist authors claim to have adopted ((Di Paolo et al., 2018, p. 138, 209). They don't want to explain the emergence of novel structure by invoking unexplained explainers. For this reason, a sense of self and other cannot be assumed to be present early in the process during which social processes complexify. Nevertheless, the participants' sensitivity to others sharpens with the increasing buildup of repertoires of partial acts and increasing regulation. It will only become more pronounced when social interaction further develops and acquires additional characteristics.

11.2.4 Linguistic Autonomy

Participatory activities grow in complexity. Their growth doesn't happen in a vacuum but in specific material and social conditions. For example, some participatory activities occur at more local scales. They might involve people who, because of kinship, shared development, and similar surroundings, already have many ways of acting, including specific sensorimotor schemes, in common. Under these circumstances, participatory activities might be regulated by specific, idiosyncratic social acts. Think about how in different cultures satisfaction with a shared meal is expressed. Local practices of regulation might not be transferrable to participatory activities that include a mix of participants living in different environments and having different backgrounds and customary ways of acting. Yet, some regulatory acts are more prone to be transferrable than others. One reason for this is their place in widely shared sensorimotor schemes, grounded in human anatomy. To avoid something or someone's continued motion, people, irrespective of local customs, can stretch out their arms with the palm facing forward to push against the object or person. In a participatory situation in need of regulation, such acts could be used without actually touching the other, as a means to intervene. In the corridor dance, it could be used to avoid the other person from choosing a side; in another context, it could be used to prevent the other person to engage—for the moment—in any participatory action. In a conversation, it could be used to allow more speaking time for oneself, because something important needs to be added before other participants reciprocate. Such acts are called "portable acts," because they can be used, with a relatively high chance of being effective, beyond the local context. Moreover, because of their grounding in widely shared anatomical and sensorimotor features, they are said to have "strong regulatory normativity" (Di Paolo et al., 2018, p. 186).

Regulatory acts create an asymmetry between regulator and regulated. Such asymmetry will become conspicuous, for example, when one participant increasingly monopolizes regulations or excessively relies on portable, strong regulatory acts. Because of such situations, the other participants will become sensitive to this particular participant as a source of regulation separate from themselves. A pointed sense of other versus oneself emerges.

Remember that, by means of regulatory acts, participatory activities are influenced. Yet, in the enactivist analysis, the situation itself retains a degree of autonomy with respect to the autonomy of the individual participants. Despite this, it is always possible that the participatory activity ceases or disintegrates. In fact, without work it will do so, because the ongoing autonomous process is precarious. Such cessation of participation might happen because of the overuse of particular portable regulatory acts by one participant. By a one-sided frequent use of portable acts, one participant may come to dominate the participation so completely that the possibility for mutual regulatory influence vanishes and, therefore, also the autonomy of the other participants and the activity as a whole. In general, for autonomous interactions to continue, the autonomy of individual participants must be maintained. A way for interactions to remain autonomous is for the participatory activity to take on a dialogic structure (Di Paolo et al., 2018, chapter 6). A dialogic structure, which is meant to apply generally to participations irrespective of whether they involve language, is a balancing of the relation between who is regulated and regulator. The regulator is the producer; whoever is regulated is the audience. A very specific way to achieve such balance is for one person to be accepted as unique producer by the other participants. Another way is for the roles of producer and audience to become distributed over time between the participants. Social acts in a dialogic structure are called "utterances." They can, but need not, take a linguistic form. Linguistic utterances emerge after repertoires of nonlinguistic utterances.

The asymmetry between producer and audience is inherent to the dialogic structure. This asymmetry refines the participants' sensitivity to the self/other distinction. Because individual participants, including oneself, become conspicuous temporary sources of asymmetric influence on the participation, a sensitivity for what they as individuals try to achieve can develop. This opens the door for a sensitivity, in participants, for aspects of a participatory act that are specific for this particular individual, and aspects that belong to the act, irrespective of the producer. A forward-facing palm might in all contexts be produced to introduce a pause, but the audience might sense that this producer uses it now because they are tired.

In other words, the possibility of interpretation emerges. Participants can disentangle the expressive aspects of an utterance (those related to the specific situation and aims of the individual producer) from the pragmatic ones (those related to the customary role of the utterance in dialogs).

Like sensorimotor schemes, utterances cluster and then cluster into larger clusters. Utterances form dialogs, and similar forms of dialoging form dialog types, that are specific to specific kinds of circumstances. These larger dialog categories are termed "participation genres" (Di Paolo et al., 2018, p. 178). Participation genres do not have to involve language, but some examples of linguistic participation genres are a meeting with friends in a café, an oral examination at the university or the shouting of slogans at a protest march. They all involve their own patterns of socially interacting. What is appropriate in one would be inappropriate in another.

Once interpretation has entered the mix, producers can apply it to the utterances they themselves produce. That is, producers can take the role of audience with respect to themselves, looking at themselves in the way others do, or the way they look at others. By this means, they can correct, replace, or amend their own utterances, after or before they are made. Self-interpretation thus is a way to self-regulate. At the same time, self-regulation is a means to regulate the dialog, as when one refrains from uttering something because one is aware it might not be correctly understood by one's audience.

A further result of the development and growth in complexity of dialogs is the use of reported utterances. Utterances made by another participant or by oneself, can be reiterated, with or without modifications, as a means to regulate the ongoing dialog. Reporting can serve a variety of purposes. For example, in a linguistic dialog, the utterance, "But you said we weren't going to talk about the weather," can redirect a dialog to its focus. On other occasions, reporting can be a means of interpreting the original producer's uttering of it, or it can be a means of specifying one's own position with respect to the utterance, to mention just a few possibilities. Reported utterances have a double nature. They "belong" both to the original producer and the reporting producer (Di Paolo et al., 2018, section 8.1.4). Di Paolo et al. (2018) describe the use of reported utterances as a form of "incorporation" (Di Paolo et al., 2018, p. 192). The reported utterance becomes part of the reporter's repertoire just as much as their own utterances, yet, retains an essential connection to its original producer.

Producers can self-report their own utterances, for example, when engaging in a process of self-interpretation. Processes of self-reporting and self-interpretation can be shielded, as much as possible, from the view of other participants (Di Paolo et al., 2018, section 9.2.2). They can take the very familiar form of silent monologues.

So, as dialogic interactions develop and diversify, structures recognizable as grammatical emerge. With an utterance that contains an utterance, recursion has emerged. Recursion, as was discussed in chapter 7, concerns the nested structure of language, allowing, among other things, for sentences that contain sentences as part of them. Crucially, in the enactive story, recursion forms first in the structure of actions. When participant B produces an utterance that recaptures and comments on A's utterance, the recursive structure is in B's reporting act. It is B's act that recaptures A's act. Recursion follows naturally from the way that interactions complexify. Unlike in Chomsky's cognitivist scheme (chapter 7, section 7.3.2) recursion doesn't enter speaking from a source "beneath" the action, such as an intrinsically recursive brain system, but emerges from within the growth of social action and interaction.

Overall, the enactivist story of language offered in *Linguistic Bodies* provides a very detailed account of the gradual buildup of ever more complex forms of action and interaction that come to be recognizable as "language as we know it" (Di Paolo et al., 2018, chapter 11). The story is a logical, not a historical reconstruction of the process leading to language (Di Paolo et al., 2018, pp. 230–231). It is meant to point out the logical order of steps which something recognizable as language needs to take to emerge from simpler forms of living and interacting. Of course, if conditions A and B are logically required for process C to happen, this logical structure needs to be reflected in the historical progression. But the alignment needn't be absolute. A logically necessary condition for process C could appear long before process C, but, of course, never after it. As mentioned, the story is explicitly presented as relying on a minimal assumptions methodology. Allegedly, at any stage in the reconstruction, nothing is assumed to be present which has not been accounted for as deriving in a plausible way from simpler conditions. Crucially, the enactive account from cell to language validates the overall enactivist life–mind continuity thesis (Thompson, 2007).

Throughout, and very much in line with the philosophers of chapter 6, the analysis points at the inherently social nature of language. Like the notion of giving a gift, the notion of utterance is essentially social. Like a gift requires someone who accepts the gift, the production of an utterance presupposes an audience to whom the utterance is addressed. Time and again, it is stressed, in Di Paolo et al. (2018) that language keeps its social character even when it is used in private or in solitary self-production (section 9.2.2). For enactivists, shared language does not derive from a preexisting individual language. It is the other way around: As for the authors of chapter 6, language is, as enactivists say, "constitutively social" (Di Paolo et al., 2018, p. 193).

11.3 The Evolution of Consciousness

Enactivism's mind–life continuity thesis concerns both consciousness and cognition. Like cognition, enactivists see consciousness as intrinsically linked with autonomous organic structure. As discussed in the previous section, Ezequiel Di Paolo has argued that some form of agency is involved in basic forms of subjectivity (Di Paolo, 2005). According to Di Paolo and colleagues, such agency can be found in early life-forms, such as single-celled bacteria (Di Paolo & Thompson, 2014; Di Paolo et al., 2018, p. 337). By linking consciousness to life, enactivists claim to achieve what according to proponents of a hard problem of consciousness cannot be done, namely, achieving an understanding of consciousness.

Other philosophers have explored the links between organismic life and consciousness, with the same aim of countering the idea that it cannot be understood for principled reasons, but without endorsing the specifics of the broader enactivist framework—built out of the concepts of autopoiesis, autonomy, or precariousness. Instead of enactivists' focus on an analysis of the structure of life, they focus on evolution and attempt to explain consciousness by seeing it as an adaptive feature acquired by natural selection. This bioevolutionary approach to consciousness has been developed in academic and broader aimed publications by philosopher Peter Godfrey-Smith (2020), and a specific version has been presented by Walter Veit (see Veit, 2023).

Overall, these biological/evolutionary philosophers criticize the enactive approach to consciousness for locating consciousness too early in the evolutionary tree. That is, they question the validity of the enactivist claim that basic forms of subjectivity arise whenever, or very soon after organic autonomy arose (Veit, 2023; section 3.3.1 targets such claims in Thompson [2007]). Further, the evolutionary approach followed by Godfrey-Smith and Veit holds that consciousness should not be thought of as a unitary all-or-nothing phenomenon, but as something that has different dimensions which all can come in different gradations. Veit, for example, distinguishes the following dimensions: self-consciousness, evaluative experience, sensory experience, the integration of experience at a time, and the integration of experience across time (Veit, 2023). He further argues that the evaluative dimension is basic. The evaluative dimension concerns attaching affective value to what one interacts with. In its crudest form, it is the capacity to experience pleasure and pain. Veit argues that having some form of evaluative experience evolved because it allowed organisms to manage how to act appropriately when behavioral complexity, and hence the spectrum of possible actions, grew (Veit, 2023).

Veit (2023) sees hedonic feelings of pleasure and pain as evolved features of organisms, and thus as traits that have an evolutionary and adaptive explanation. Hedonic feelings serve as a "common currency" for the organism, allowing to *feel* the value of different behavioral options and, based on these feelings, to choose between them. Feeling thus is a functional biological trait: It makes a substantial difference to be an organism that has hedonic feelings. Once basic hedonic feeling has evolved, biological evolution could find ways in which it could not have gone without that evolutionary innovation.

Veit further defends that hedonic feeling is the most basic dimension of consciousness. The other dimensions of consciousness evolved only once it was in place. Thus, it is possible for an organism to have forms of sensory sensitivity—it might detect and show behavioral reactions to sensorimotor contingencies—without actual or potential awareness. That is, the organism would never become aware of its sensory interactions if it had not evolved evaluative sensitivities. But if an organism has evolved a capacity for hedonic feeling, this typically becomes integrated with its sensory capacities. Sensing and perceiving then is both world-relating and evaluative: We taste sugar—the perceptual aspect—and experience sweetness—the hedonic aspect—at the same time. What goes for the sensory dimension of consciousness goes for the the other dimensions. The sense of self and awareness of time only become dimensions of consciousness when they are integrated with evaluative consciousness (Veit, 2023).

11.4 Different Enactivisms

Enactivism, since it was first formulated by Varela et al. (1991), has diversified. There are now different positions that are known as "enactivism." The first is the one originating from the work of Varela and coauthors. It is known as "autopoietic," "autonomous," or "original" enactivism (Di Paolo & Thompson, 2014; Hutto & Myin, 2013; Käufer & Chemero, 2015). The second is "sensorimotor enactivism," associated mainly with the influential paper of O'Regan and Noë (2001). As discussed, it was meant to offer an answer to the "hard problem" of accounting for perceptual consciousness. Seeing perception as something we enact or do, that is, as an activity of persons in environments, structured by sensorimotor contingencies, dissolves the hard problem, so sensorimotor theorists claimed. Third, there is "radical enactivism," as deriving from the work of Hutto and Myin (Hutto, 2005; Hutto & Myin, 2013, 2017).

In two books, Hutto and Myin defend a position they called "REC," short for "radically enactive/embodied account of cognition" (Hutto & Myin, 2013, 2017). Hutto and Myin's overall position is that thinking about the mind, both in science and in philosophy, should abandon cognitivism and its allied philosophy. The emerging fields of embodied, dynamical, and enactive thinking, in both the original and sensorimotor version, contain crucial ingredients to replace the internalist, computationalist, and representationalist approach to the mind. Yet, embodied and enactive approaches should break radically with the cognitivist tenets, and not partially conserve the old framework, for example, by replacing mental representations with embodied representations (as in chapter 10, section 10.1).

Hutto and Myin argue that a radical embodied/enactive position should overcome the foundational idea of standard approaches to the mind that cognition is always a matter of contentful representation or contentful information processing. As we have seen, the idea of the mind as essentially representational was already present in ancient Greece, and later it formed the heart of what Reid called "the way of ideas." This approach was taken for granted in much scientific psychology and was reaffirmed with vigor in cognitivist-style thinking about the mind. As we have seen in chapters 7, 8 and 9, in philosophy of mind representationalism has become the default way of thinking about many mental phenomena, including perception.

Hutto and Myin put into question the foundational assumption that cognition always and everywhere involves representational content. Content captures the intuitive idea of meaning. In philosophy, it was this credo that led to analyzing even basic cognitive capacities, such as perception, in contentful terms. In the sciences of the mind, this assumption appeared as the conviction that mind and brain were essentially processors of information, or as essentially engaged in performing computations on representations.

As explicated in chapter 9, contentful representation involves some kind of specification, picturing or describing of something that is represented by a representation. What is represented by a representation is the representation's content. Content is often defined in terms of truth conditions—or the very similar notion of accuracy conditions. A representation's content is a specification of what would be the case if the representation were true. Therefore, the meaning of "It will rain in Monaco this afternoon" has the content that it will rain today in the afternoon in Monaco.

The problem with reliance on the assumption that cognition is inherently contentful, so Hutto and Myin's REC claims, is that it should be explained how this inherent, original, and ubiquitous content comes to be. For example, how is it possible that perception, a capacity shared by many animals and present in young human infants, can carry contents? Adult humans have a firm grip on what it means for a sentence such as "It is raining" to represent in a truth-conditional way. This much is endorsed by Hutto and Myin, but they see the sentential example as standing in contrast to "perceptual states." For if it is claimed that "naked" perception, as when a dog perceives a cat in front of a tree, is a contentful state, how has this state acquired its capacity to represent? Contrast this with the spoken claim, "It rains today." Whoever makes such claims has acquired this capacity via an extended developmental process. During this process, the person gradually learned "the means to mean," to construct and utter descriptions. Along the way, this person has acquired a sensitivity for truth, they learned to participate in truth-telling practices. They learned what kinds of responsibilities one takes on when one makes a statement—to be able to provide evidence, to retract one's claim if good reasons appear that cast doubt on it, and so on. Moreover, such truth-telling practices have had to develop themselves in a sociocultural process during which social interaction increasingly complexified, possibly along the lines analyzed in *Linguistic Bodies* (Di Paolo et al., 2018). REC claims that it makes no sense to think of mental phenomena as carrying truth-conditional content except against the background of truth-telling practices.

Philosophers who endorse what is now known as "the content view of perception," a view attacked by Hutto and Myin, often point to philosophical theories of content, discussed in chapter 9, as justifying their position. They hold that such theories show how, among other cognitive states, perceptual states, contra REC's claims, can carry content. The theory of content that is most relied on as allegedly providing a foundation for the content view is teleosemantics à la Millikan (see chapter 9). Roughly, under such analysis, a perceptual state, or a neural state for that matter, of an organism carries content about rain if two conditions are met. First, it has been selected by natural evolution to react specifically to rain. Second, it has been selected to control other mechanisms underlying the organism's behavior with respect to rain.

Hutto and Myin agree with teleosemantics that evolution can select for specificity. Under selective evolutionary pressure, organisms can acquire capacities to selectively respond to and engage in specific activities with respect to specific objects, events, or aspects of the environment. But, according to REC, such selective responding and functioning does not amount to representing. Evolution accounts for being specific, but that is something different than accounting for specifying. A safety fuse for the electric circuit of a garden will be specifically sensitive to moisture and have a specific influence on the circuit when moisture is detected, but it doesn't thereby represent moisture, rain, or whatever.

Hutto and Myin (2013) stress that, with respect to cognition, the distinction between intentionality—intentionality-with-a-t—and intensionality—intensionality-with-an-s—is crucial. Intentionality is directedness; intensionality is directedness under a description. Intensionality is always representational; intentionality isn't. My car can be sensitive to the rain by switching on the window wipers; I can be sensitive to the rain by conceiving of it as "the rain," as "what spoils the weekend plans" or as "what finally gives the garden zucchinis what they desperately need." REC agrees with the older criticism of teleosemantics by Jerry Fodor (see chapter 9). Fodor used the example of a teleosemantic treatment of neurons in the frog's brain involved in detecting flies and initiating subsequent motor reactions. On a teleosemantic account, the neurons would be representations carrying a content like "there's a fly in front of me." Fodor complained that the way the content was described—its intensional guise—was arbitrary. Nothing in the functioning of the frog, or in its evolutionary history, fixes which description of the flies is the correct one (Fodor, 1990). Hence Fodor quipped: "Darwin doesn't care how you describe the intentional objects of frog snaps. All that matters for selection is how many flies the frog manages to ingest in consequence of its snapping" (Fodor, 1990, pp. 72–73).

In other words, the complaint, shared by Fodor and Hutto and Myin, is that biological function, the axis around which teleosemantics revolves, even if it would explain intentionality-with-a-t, does nothing to explain intensionality-with-an-s. But if representation is always intensional, as is standardly assumed, then teleosemantics falls short as a theory that accounts for representation in terms of evolutionarily established function.

Hence, for Hutto and Myin, basic perception is intentional, but not intensional. Evolution has equipped basic perceivers with abilities to be perceptually in touch with their environments. Basic perceivers see trees, or food, and enjoy conscious experiences while doing so, but they don't thereby perceptually represent trees or food in some descriptive, contentful way. Their perceptual experience itself is not true or false. Basic perception is specific, but it doesn't specify. Perceptual experience can go wrong, as when one occasionally experiences what is specific for cat circumstances in dog circumstances, but this amounts to no more than a relation to biological, adaptive norms, instead of norms of truth and falsity.

How should intensionality, and thereby representation, then be explained, according to Hutto and Myin? When and how does intentionality become intensionality? What is needed, they state, is what they call an account of "the natural origins of content" (Hutto & Satne, 2015; Hutto & Myin, 2017, chapter 6). This means: a developmental story of the phylogenetic and ontogenetic conditions under which intensional representing emerged. Such is a story about how genuine representers acquired the means to mean by engaging in content-involving activities, such as saying "It rains today" upon looking through the window. So Hutto and Myin don't deny the existence of fully fledged truth-conditional, intensional content. They do claim that such content, as far as humans are concerned, arose as a consequence of the establishment of socio-culturally shared truth-telling practices.

How does REC relate to original enactivism? REC endorses that, for all we know, life offers the best model to allow for an understanding of mind (Hutto & Myin, 2013, p. 4). It also endorses the importance of history, and histories of interaction in such understanding of the mind. Yet, while endorsing biology and evolution, REC has remained neutral with respect to core concepts or original enactivism such as autonomy. REC has been critical of the language of "sense-making", and its application to phylogenetically early life-forms, such as single-celled bacteria (Hutto & Myin, 2013, chapter 2; 2017, pp. 75–82). The notion of making sense seems to imply some kind of specifying, picturing or describing that is constitutive of the notion of content. If sense-making is not truth-conditional content (or something closely related), it has to be explicated what this different form of specifying, picturing or describing is and why it is a specifying that is done, sense that is made, *by* the organism.

11.5 Evaluating Embodied and Enactive Thinking

In this chapter and the previous one, we have dealt with embodied, embedded, and enactive approaches. Though it is important to remain aware that there are many differences between and within such positions, most of them share characteristics. Many of them oppose a similar set of views about the mind that, as has been discussed in this book, have achieved mainstream acceptance. To a large extent, they oppose cognitivism and its core assumptions, as well as the idea that consciousness presents an unsolvable hard problem, because of the existence of unexplainable "qualia" (or something like it). For simplicity's sake, let's from now on use the term "E-thinking" for all those embodied, embedded, and enactive approaches that reject cognitivism. These include Gibsonian ecological psychology, many embodied cognition positions, and most dynamical systems and enactive accounts.

As we have seen, E-thinking on consciousness tends to affirm that we can understand conscious experience, by connecting it to where it comes from; to its material, biological embedding and to how it is "enacted" in concrete circumstances. Taking this stance means rejecting that there are a priori or in principle reasons—very general arguments deriving from alleged insights in reality or logic—that force us to conclude that no explanation or understanding of consciousness can be obtained in a broadly naturalist, scientific perspective. Though E-thinking about consciousness is in many ways unique and novel, it resonates with how other thinkers have looked upon consciousness. The idea that sensory consciousness is widespread in the biological world was an integral part of Aristotle's philosophy of nature. Resistance to a priori arguments for a special problem of consciousness and phenomenal qualities can be found in Carnap and the identity theorists. Wittgenstein and Anscombe pointed to life and living as belonging to the circumstances that needed to be invoked to gain a genuine intellectual grasp on all aspects of minds.

With respect to cognition, the same contours are visible. Thinkers endorsing the core assumptions of cognitivism have advanced many arguments for why forms of E-thinking that reject cognitivism *must* be misguided or of minor significance. For example, they have argued that E-accounts of cognition are at best limited to what they call low-level cognition (Clark & Toribio, 1994). Low-level cognition, as when one sees and grasps an object in front of one, is interactive: It concerns what is present in the situation now. But higher forms of cognition rely on what's not present in the interactive situation: on what one remembers or knows, for example. To act in the present in light of what happened in the past, one must rely on stored knowledge or mental representations of the past. Anything but perhaps the most simple or basic forms of cognition requires representation as understood in the cognitivist framework. Things only get worse for E-thinking when one considers that cognition is abstract and generative. We interact with concrete objects—an apple, for example—but the way we interact with it is shaped by the abstract category the object we interact with is taken to belong to. And, so the reasoning continues, taking something as an instance of an abstract category requires mentally representing it as such (Clark & Toribio, 1994). Further, though each of the interactions in which intelligence or cognition is displayed, is a concrete action, the totality of these interactions shows a structure. How someone acts, perceives, thinks, and speaks now is systematically related to how they acted, perceived, thought, and spoke before and how they will act, perceive, think, and speak in the future. To explain this systematicity or generativity, the existence of a producing or generating source, that is itself systematic, must be assumed (chapter 7, section 7.3.3). Because E-thinkers focus only on the interactions, such behind-the-scenes generative systems must fall outside of their scope. In short, E-thinking is criticized for seriously overestimating interaction per se and therefore massively underestimating the necessary complexity of what is needed to understand and explain cognition. The reasoning leading to this conclusion can be discerned in Chomsky's and Fodor's approaches, but it is also shown some sympathy for by Dennett (Dennett 1987; see chapter 8, section 8.5). It is present more generally in different kinds of computationalism, because computation is considered to be a generative process (see chapters 7 and 8).

E-thinkers reject such arguments. They protest that it is mistaken to portray their focus on interaction as a focus on cognition that is low-level and cannot involve what's absent and what's abstract. Consider someone heading to the station to take a train to travel to a city where they are having a job interview. The person is interacting with the environment in the here and now. At one moment, they swing one leg before the other. But the swinging of the leg is at the same time a bodily movement, a walking to the train, and the unfolding of a plan to apply for a position (van Dijk & Withagen, 2016). The fact that low-level and high-level cognition cannot be separated here is reflected in the nestedness of spatiotemporal scales in dynamical systems. What a cognitive agent does in a small temporal window is nested in and codetermined by what they do on wider timescales, akin to how cell division in an embryo is nested in and determined by the whole somatic developmental process. In construing cognition as dynamically unfolding at multiple coordinated temporal and spatial scales, low-level, high-level, and offline cognition are seen as inseparably integrated or meshed—the distinction makes sense only as an abstraction, not as something that has concrete counterparts in the situation. Crucially, E-theorists endorse that the absent train plays a role in the current activity of walking to the station. But the story they offer for how that role is played differs from a mediationist narrative according to which the absent train can only influence activity if it is present in the actor as a mental representation. E-theorists hold that the reason why the activity of walking to the station concerns the absent train, are spread across space and through time, instead of being distilled in a punctuated, causally active mental representation in the brain. The person walking to the station has become, in an extended temporal process, familiar with trains and how to act with respect to them. They know what to do when they find themself in a situation where a train forms the best means to reach a distinct destination and can act accordingly. They might imagine the train, or consider alternative routes before setting off, but imagining and considering are embodied, embedded, and enacted activities too.

In the same vein, E-theorists do not restrict themselves to cognitive activity that is not abstract. They don't deny that cognition is often abstract, but they will claim that abstraction should be explained in interactive, dynamical terms. Abstraction is acquired; it is a product of a history of interactions, rather than a basis. E-theorists will treat the generative or systematic nature of cognition in the same way. Generativity, for them, is a result too. It is something that has evolved along the development of the many varieties of cognition. In fact, the enactivist account of language surveyed in section 11.2 of this chapter provides the outlines of a story of how the generativity of language can arise out of the complexification of social actions. Crucially, this explanation of generativity refers to specific material, bodily, environmental and social conditions. "Language as we know it" (Di Paolo et al., 2018, chapter 11) is generative because it has become so in a historically stretched out process that involves life, bodies, environments, social constellations and artifacts. E-theorists will point out that there's no way around invoking this spatiotemporal entanglement in one's account of mind and cognition. Complex phenomena require complex explanations.

Bibliographical Note for Chapter 11

An important and influential work on enactivism is Evan Thompson's *Mind in Life: Biology, Phenomenology, and the Sciences of Mind* (Thompson, 2007). As the subtitle says, its version of enactivism is very much inspired by philosophical phenomenology. Another influential author on enactivism is Shaun Gallagher; see, for example, his *Enactivist Interventions. Rethinking the Mind* (Gallagher, 2017). Enactivism has been applied to other areas than language. See, for example, Sanneke de Haan's *Enactive Psychiatry* (2020).

The idea that grammar arises "along the way" out of the complexification of social life has been defended in linguistics too, and in the cultural evolution literature. See, for example, the work of Michael Tomasello (e.g., Tomasello, 2005) and of Joseph Henrich (Henrich, 2016). Tomasello stays much closer to the cognitivist tradition than E-theorists. But not close to Chomsky, as can be seen from Ibbotson and Tomasello (2016).

A unique perspective on the evolution of mind and eventually consciousness is provided by Fred Keijzer's "skin–brain hypothesis." According to Keijzer, nervous systems evolved initially as action-coordination systems, and not for transforming inputs into outputs. Perception is a result of, not the basis of action coordination (Keijzer, 2015).

12 Epilogue: Forever in the Loop

> The end has no end.
>
> —The Strokes, *Room on Fire*

Ultimately, what we have been dealing with in this book are attempts to come to grips with who we are. Philosophical and scientific thinking about the mind aims at understanding the mind. But because the mind that thinks is also the mind that is the target of understanding, this thinking is always both about the mind that it tries to understand and the mind that does the understanding.

Unlike rivers and stones, minds are both object and subject. Of course, we can study the mind, and the brain and body that are, according to most philosophies, tightly connected to it, as objects. We can talk about thoughts or feelings as going on inside us, as having an imaginary unicorn before the mind's eye, as seeing the film of our lives passing by. And we can, with increasing precision, observe what happens in our brains and bodies when we think, imagine, or perceive.

But even when we think about our thinking in terms of thoughts, or of our imagination as a picture, we are doing something: thinking. We are actively inside this thinking as the person that does the thinking. We animate the thought, and when we think it, it is active. The original thought becomes objectlike only when we think a new thought about it, that takes it as object. But then our new thought is active, and not objectlike. What goes for thinking and thought goes for perceiving and feeling. Talking about a perception and a feeling can be useful, but it should never be forgotten that there is a perceiver of perceptions, and a feeler of feelings.

How can minds be both objects and subjects? One possibility that has tempted thinkers is to assume the existence of two kinds of facts or properties: objective facts and properties and subjective facts and properties. This separation forms the core of Descartes's substance dualism, and one can still discern it in some of the contemporary literature on qualia and phenomenal properties.

But is positing a self-standing category of subjective facts not simply providing a label for the problem instead of a solution to it? Another idea for dealing with the double-faced nature of minds is to insist that circumstances and surroundings, forms of life, life itself, and action are crucial for understanding the mind. Subjects exist because agents exist, and agents exist because life exists. In chapters 4, 6, 10, and 11, we have seen varieties of such strategy—with significant differences between them. What all these action and life-based approaches have in common is that, one way or another, they stress the need to take a broad view when one tries to come to grips with minds. To understand minds, one needs to not just look at what's currently happening in the brain, but also at the body and how it is situated in its environmental and social situation. Further, history matters. Minds have evolved, and understanding precisely how they have evolved might make us understand what we found impossible to grasp before.

To come to understand something is often to see connections one didn't grasp before. The instructions for assembling your freshly bought cupboard might leave you feeling helpless till you find out how to match actions to drawings and numbers. Your eventual understanding is forged by actively connecting perceiving and doing. A back-and-forth between looking and doing ensues in which the doing shapes the seeing and the seeing the doing. In the end, all the pieces connect, literally and figuratively.

Perhaps things are not that different when it comes to understanding minds. Maybe acquiring a grip on seemingly intractable problems about the mind should be viewed on the model of the cupboard example. It takes shape in the act of finding the right connections in the wide and growing body of knowledge about how organisms act, interact, and how this all has come about.

Consider again the double-faced nature of minds. This aspect of minds might become more tractable when we notice that it arose in evolution once some form of agency emerged. Once organisms with agency evolved, these agents started acting in ways that determined their further evolution. Just by changing their environment, they change the course of evolution. Natural selection becomes partly under their own control. Evolved agents become both results and causes of their own evolution, both objects and subjects.

Coming to understand the mind might require us to rethink what it is to understand the mind. But now a further intricacy arises. For if we are both subject and object, changing our understanding of the mind changes the very minds that are understood. And that process continues as comprehension grows. To make progress in understanding the mind, there might be no other choice than to become of a different mind.

Bibliographical Note for Chapter 12

I finish with a note on two philosophers who, throughout this book, have not been given as much explicit attention as they deserve. Maurice Merleau-Ponty's classic *Phenomenology of Perception* has many insightful passages on the double, subject-object, nature of perceivers. (Merleau-Ponty, 1945/2012). Susan Hurley's *Consciousness in Action* remains an outstanding resource for advancing thinking on the nature of action, mind, and self (Hurley, 1998).

Glossary

Note: This glossary is meant to briefly explain crucial terms as they are used in the book. When the explication of a term mentions one of the other glossary terms, these terms are printed in italic. Examples of types are also in italic, but that should cause no confusion.

adverbialism: To claim that the appropriate use of a word should be modeled on the use of an adverb, even if the word isn't an adverb. So, to say that someone does something with intelligence, means, on an adverbialist reading, that someone does something intelligently, in an intelligent way. The adverbialist rendering can be seen as a means to avoid the *reification* of intelligence.

affordance: Often used in the plural and referring to the action possibilities provided by something in an organism's environment. The term is central in J. J. Gibson's ecological approach to visual perception. Gibson claimed that organism's *directly perceive* affordances. This means that a human would immediately see the affordance "allowing to be sat upon" when they look at a chair. They, or their visual system, do not see the chair first in a neutral way to then infer or add that it can be sat upon.

ambient optic array: Terms in J. J. Gibson's ecological approach to visual perception for the structure of the light that is available at any point in space and that allows organisms to see what's in the environment. Thomas Reid's notion of "visible figure" is quite similar.

analytic: A sentence is analytic if its truth or falsity depends only on the meanings of its terms. So, the expression "Red is a color" is analytically true, because it is true only because of the meanings of the terms "red" and "color." "All points on a circle are at the same distance from its midpoint" is analytically true because its truth depends only on the definition of a circle as "the figure consisting of all the points equidistant from a given point."

a priori: Regarding knowledge, what one can know without checking the facts. Often, analytic sentences such as "Red is a color" are taken to be knowable a priori. It is standardly held that mathematical and logical truths are analytic. "A priori" is opposed to "a posteriori."

a posteriori: Pertaining to knowledge one knows only after one has checked the facts. For example, how much you weigh now, down to the ounce, is knowable only a posteriori. It might no longer be possible to check the relevant facts. How much Aristotle weighted at his birth is still a piece of a posteriori knowledge. Opposed to *a priori*. *Synthetic* sentences are generally considered to express a posteriori knowledge.

attractor: A global condition of a dynamical system to which it in certain circumstances tends to evolve.

autonomy: Generalization of *autopoiesis* beyond biomolecular systems. Autonomous systems self-generate, but always in precarious conditions.

autopoiesis: A biomolecular system whose constituent processes generate and recreate each other and, thereby, the system, exemplified by a living cell.

behaviorism: Family of a variety of views that in one way or another prioritize behavior or action when it comes to understanding the mind.

cognitivism: An approach in the scientific study of the mind that takes *computation* and *representation* to be key explanatory concepts. Cognitivism became dominant after the so-called cognitive revolution, roughly midway through the twentieth century. There used to be a strong tendency towards *internalism* in cognitivist theorizing, but this aspect can be relaxed.

computation: Generalization from the idea of calculation. Roughly, the transformation of formulas built from the characters of some alphabet of characters into other formulas built from characters of the same alphabet. There are detailed mathematical theories of computation.

connectionism: An approach in cognitive science in which neural network models are made to explain cognitive phenomena. Neural networks consist of units—idealized neurons—that are connected to each other. Neurons have a degree of activity and can influence the activity of other neurons they are connected with. Learning is modeled as changes in the connections, and knowledge as patterns of activity in the network.

content: Generalization of the idea of meaning of a linguistic expression. Meaning can be thought of here as what a sentence says or describes. Many philosophers think there are also different kinds of content than the meanings of expressions in language. They think there are mental representations that carry such content. Like meaning, content requires that the content of a representation can or cannot align with reality. It can be true or accurate, or false or inaccurate. Content thus requires a means to mean, a medium for expressing content. Philosophers have tried to formulate theories of content to account for the content of mental representations. *Teleosemantics* is a prominent theory of content.

direct perception: This phrase can have two meanings. It can refer to the idea that the environment is perceived without something else being perceived first, consciously or unconsciously. Direct perception is then opposed to *mediationism*. In another sense, perception can be said to be direct if it does not rely on inferences, conscious or unconscious.

disposition: A tendency to do certain things or to show certain reactions or behaviors in certain circumstances. Glass has the disposition to break when it falls to the ground. People in pain have the disposition to shout.

dualism: The thesis that there is a fundamental difference between the mental and the physical. Often a distinction is made between substance dualism and property dualism. Descartes's dualism is a substance dualism as it posits two different substances. A property dualist needn't go that far, but any property dualist will make the claim that there is a fundamental difference between mental and physical properties.

embodied cognition: A theoretical approach in the sciences and philosophy of the mind in which the body is assumed to play an important role for cognition. Some forms of the embodied approach stay true to computationalism and representationalism (see *representation* and *computation*). Other embodied cognition theorists see embodied cognition as opposed to cognitivism, because they think cognition should be explained in terms of embodied activities rather than of in terms of computations and representations.

empiricism: Approaches to knowledge and mind that hold that, in some way, all knowledge derives from what is acquired by the senses.

epiphenomenalism: Things or properties that are caused by a process but that do not influence that process themselves.

idea: In the Platonic tradition, an idea is an abstract item that forms a model for particular things and is more real than particular things. Concrete round things are based on the idea of a circle, which is abstract. In the tradition of modern philosophy, ideas stand for the constituents of mind, ranging from perceptions to thoughts and feelings. This use of "idea," or something close to it, can be found in Descartes, Locke, and Hume, for example.

idealism: The idea that material objects do not exist independently from minds but are constructions out of ideas. Idealists take ideas to be mental and therefore hold that everything is mental. George Berkeley (1685–1753) was a famous idealist philosopher.

indirect perception: The idea that we don't perceive the world directly but only by intermediacy of images, representations or models of the world.

information: In a narrow sense, information is related to causation or correlation. In this sense effects can be said to carry information about their causes. Smoke carries information about fire. In a broader sense, information is related to meaning. The sentences in a book carry information, or a whole book might contain information. Some philosophers have tried to argue that one can understand information in the broader sense as deriving from information in the narrow sense.

Information processing: Performing operations on *information*. When people talk about information processing, it is often not indicated whether they mean information in the correlational or in the meaning-related sense.

internalism: Regarding cognition, the thesis that the essential aspects of cognition happen within some boundary, mostly the brain or computational system.

intensionality: When it matters not only what is represented but how it is represented, intensionality comes in play. Intensionality exists when there are different ways of representing the same reality. One can represent Shakespeare as "The author of *Hamlet*" but also as "The author of *Romeo and Juliet*." These two expressions have the same *reference* or extension, but a different *sense* or intension. Some forms of knowing show intensionality, because one can know that "Shakespeare wrote *Hamlet*" without knowing that "The author of *Hamlet* was the author of *Romeo and Juliet*."

intentionality: The directedness of the mind or language. A thought is intentionally directed upon what the thought is about. A fear is intentionally directed upon a thing that is feared and so on. Descriptive sentences in a language are directed upon the things in the world they describe.

knowing-how: Knowledge of how to do things, for example, tying your shoelaces. One has knowledge how of something if one can do it, it is not required that one can describe how one does it.

knowing-that: Knowledge of facts or descriptions. Knowledge that can be reported. If one has knowledge that Paris is the capital of France, one can say so.

mediationism: With respect to perception, the claim that perception does not directly reach the world, but only through something that mediates between the world and the perceiver.

modern synthesis: The combination of the idea of Darwinian evolution by natural selection with Mendelian genetics. Darwin had no theory of how heredity worked, and that was provided by Mendelian genetics. The modern synthesis came about roughly halfway the twentieth century and was further articulated when the role of DNA in genetics became known.

monism: The idea that reality is fundamentally one, that reality is not split up in different kinds of realities. Monism is opposed to *dualism*.

naturalism: The position that everything is "natural," part of one nature. Naturalists strive to acquire an understanding of the whole of reality that does not include something supernatural, something standing outside nature. Some naturalists are physicalists, but *physicalism* is often more restrictive than naturalism. Sometimes physicalists commit to specific views that naturalists don't need to be committed to. A physicalist could hold that all that exists or all that happens can in principle be described by the science of physics, but a naturalist needn't agree. A naturalist could, for example, accept the natural emergence of properties in biology out of chemical and physical properties. Novelty can arise within nature, according to some forms of naturalism.

ontogenetic: Term in biology to refer to the timescale on which an individual develops. Growing and learning are processes that occur on the ontogenetic timescale.

panpsychism: The thesis that everything in reality—including objects like stones—has mental properties.

phenomenal: The conscious properties of something. Seeing a strawberry has physical aspects, involving light, activity in the eye, optical nerve, and brain, but it also has phenomenal properties. Sometimes the use of the term "phenomenal properties" is very similar to the use of "qualitative properties," or *qualia*.

phenomenology: In its broadest sense, this just means "what happens." In the context of the mind, it often is used for "the conscious aspects of what happens in the mind." The phenomenology of pain would concern how the pain is felt by a subject. The word is also used for a philosophical approach to the mind that ascribes great importance to phenomenology in this second sense. Edmond Husserl (1858–1938) is the founder of this philosophical school.

phylogenetic: Term in biology to refer to the timescale on which different species develop.

physicalism: The thesis that everything that exists is physical. Sometimes physicalism is tied to commitment to a special view of "the physical," for example, what is studied by physics.

proposition: Philosophical term for the meaning of a sentence. "It rains" and "*Il pleut*" express the same proposition in English and French. What exactly a proposition is, is a topic of philosophical discussion, and answers vary.

propositional attitude: Believing that it rains, hoping that the train will arrive, or fearing that the night will be cold are all propositional attitudes. These expressions contain a verb denoting an attitude (believing, hoping, and fearing) and a descriptive sentence, whose meaning is often termed a *proposition*.

quale/qualia: The conscious aspects of experience, very similar to *phenomenal* properties, but arguably not every phenomenal property is a quale. If someone experiences red, there are many phenomenal properties to the experience. It is personal and comes with a specific felt temporality, for example. But the experience's quale would be taken to be "the redness of red." So, the use of quale language often goes hand in hand with the assumption that there is some specific characteristic property for that kind of experience. Some philosophers have argued that the notion of a "quale" does not make sense and should be dumped. They do not need to deny the existence of consciousness to make this dismissive move. Their critique concerns a concept for experience, not experience itself.

reference: What is meant by a word, sentence, or representation independent of how it is described or represented. "The author of *Hamlet*" and "The author of *Romeo and Juliet*" and "*L'écrivain Shakespeare*" all refer to the same person, Shakespeare, but represent him in a different way. These expressions have the same reference but a different *sense.*

reification: Making into a thing that what is not a thing. To use an example from Ryle: to think that "team spirit" is a tangible thing that exists in itself apart from the activities of a team.

representationalism: The idea that representations play a fundamental role in cognition. Representations are often conceived of as some kind of descriptions, pictures, images, or characterizations. Representation then carry *content.*

sense: In philosophy of language, a way of describing or representing. Two descriptions can have the same reference, but a different sense. For example, both "The author of *Hamlet*" and "The author of *Romeo and Juliet*" refer to the same person but under a different mode of representation, or sense.

sensation: Though the word has many meanings, in thinking about the mind, sensations are often taken to be simple experiences, like the experience of a particular tone or a particular color. Some philosophers and psychologists have thought of sensations as building blocks of more complex experiences.

substance: Philosophical term stemming originally from the Aristotelian tradition. A substance is a thing that can exist on its own. In the Aristotelian tradition, any object was analyzed in terms of a substance and forms. The substance was thought of as the unshaped material and the forms then "shaped" it so that it had properties. In Descartes's work, the notion of substance comes to be used for a division between fundamentally different kinds of reality. This use is quite different from the Aristotelian one.

synthetic: An expression whose truth or falsity depend on the facts. "It will rain tomorrow" is a synthetic expression, because we'll have to wait and see if it will rain the next day.

teleosemantics: A philosophical theory of *content.* Teleosemantics claims that some neural, mental, or linguistic structures carry content because of the combination of two factors. First, the structures were selected to be sensitive to certain things or aspects of the environment. Second, the structures provide input to other processes that lead to action with respect to what the structures are sensitive to. In this way, teleosemantics aims to account for intentionality, or how the minds of organisms can be directed toward specific things and aspects of its environment. Whether teleosemantics accounts for *intensionality*-with-an-s is contentious.

token: An individual that belongs to a certain general *type*. The concrete individual dog "Fido" is a token of the type dog. Tokens are sometimes said to realize or instantiate a *type*.

topic-neutral: Jack Smart's term to make the claim that our common mental vocabulary does not imply a commitment to nonphysical properties or entities. Our mental vocabulary leaves it open whether or not the mental is physical.

Turing machine: A mathematical model of a computing machine that is taken to be able to compute everything that can be computed.

type: Something general. Plato's ideas are types, but not every contemporary thinker needs to agree with Plato on what it is for something general to exist. Types can be realized or instantiated by *tokens*. The type *tree* can be instantiated by any particular tree.

References

Adamson, P. (2014). *Classical philosophy: A history of philosophy without any gaps* (vol. 1). Oxford University Press.

Adamson, P. (2016). *Philosophy in the Islamic world: A history of philosophy without any gaps* (vol. 3). Oxford University Press.

Adamson, P. (2019). *Medieval philosophy: A history of philosophy without any gaps* (vol. 4). Oxford University Press.

Akins, K. (1993). What is it like to be boring and myopic? In B. Dahlbom (Ed.), *Dennett and his critics. Demystifying mind* (pp. 124–160). Wiley-Blackwell.

Albright, T. D. (2015). Perceiving. *Daedalus, 144*(1), 22–41.

Anscombe, G. E. (1957). *Intention.* Basil Blackwell.

Anscombe, G. E. (1958). Modern moral philosophy. *Philosophy, 53,* 1–19.

Anscombe, G. E. (1983). The causation of action. In C. Ginet (Ed.), *Knowledge and mind* (pp. 174–190). Oxford University Press.

Aristotle, C. (350 BC/2011). *Peri Psychēs/De anima/On the soul.* Translated as *On the soul* by Mark Shiffman. Focus.

Avicenna, C. (1027/ 1955–1958). *Danesh-e Nâma/Livre de Science* (vols. 1–2, Mohammed Achena and Henri Massé, Trans.). Les Belles Lettres. Paris.

Armstrong, D. M. (1968). *A materialist theory of the mind.* Routledge.

Ayer, A. J. (1936). *Language, truth and logic.* Victor Gollancz.

Bäck, A. (2008). Aristotle's abstract ontology. In *The society for ancient Greek philosophy Newsletter* (p. 377). Birmingham University Library.

Bailey, S. (1842). *A review of Berkeley's theory of vision: Designed to show the unsoundness of that celebrated speculation.* James Ridgway.

Barrett, L. (2011). *Beyond the brain. How body and environment shape animal and human minds*. Princeton University Press.

Barrett, L. (2012). Why behaviorism isn't satanism. In J. Vonk & T. K. Shackelford (Eds.), *The Oxford handbook of comparative evolutionary psychology* (pp. 17–38). Oxford University Press.

Barrett, L. (2016). Why brains are not computers, why behaviorism is not satanism, and why dolphins are not aquatic apes. *The Behavior Analyst, 39*, 9–23.

Barsalou, L. W. (1999). Perceptual symbol systems. *Behavioral and Brain Sciences, 22*, 577–609.

Baum, W. M. (1998/2005). *Understanding behaviorism. Behavior, culture and evolution* (2nd ed.). Blackwell Publishing.

Bayne, Tim. (2022). *Philosophy of mind: An introduction*. Routledge.

Ben-Zeev, A. (1986). Reid's direct approach to perception. *Studies in History and Philosophy of Science Part A, 17*(1), 99–114.

Berkeley, G. (1710). *A treatise concerning the principles of human knowledge*. Aaron Rhames.

Berwick, R. C., & Chomsky, N. (2017). *Why only us: Language and evolution*. MIT Press.

Blau, J. J. C., & Wagman, J. B. (2023). *Introduction to ecological psychology: A lawful approach to perceiving, acting, and cognizing*. Routledge.

Block, N. (1978). Troubles with functionalism. In C. Wade Savage (Ed.), *Perception and cognition: Minnesota studies in the philosophy of science* (vol. IX, pp. 261–325). University of Minnesota Press.

Boyd, R. (2018). *A different kind of animal: How culture transformed our species*. Princeton University Press.

Boyd, R., & Richerson, P. J. (1985). *Culture and the evolutionary process*. University of Chicago Press.

Brentano, F. (1874/1973). *Psychologie vom empirischen Standpunkt*. Duncker & Humblot.

Brewer, B. (2011). *Perception and its objects*. Oxford University Press.

Brooks, R. A. (1991). Intelligence without representation. *Artificial intelligence, 47*(1–3), 139–159.

Bruineberg, J., Withagen, R., & Ludger van Dijk, L. (2023). Productive pluralism: The coming of age of ecological psychology. *Psychological Review, 131*(4), 993–1006.

Buchanan, B. G., & Shortliffe, E. H. (Eds.). (1984). *Rule-based expert systems: The MYCIN experiments of the Stanford Heuristic programming project*. Addison Wesley.

Buckner, C. (2019). Deep learning: A philosophical introduction. *Philosophy Compass, 14*(10), e12625.

Buckner, C. (2024). *From deep learning to rational machines: What the history of philosophy can teach us about the future of artificial intelligence*. Oxford University Press.

Carnap, R. (1931/1959). Überwindung der Metaphysik durch logische Analyse der Sprache. *Erkenntnis, 2*, 220–241. (Reprinted as The elimination of metaphysics through logical analysis of language. (G. Shick trans.). In A. J. Ayer (Ed.), *Logical positivism* (pp. 60–81). Simon and Schuster.

Carnap, R. (1933/1959). Psychologie in physikalischer Sprache. *Erkenntnis, 3*, 102–142. (Reprinted as Psychology in physical language. (G. Shick trans.). In A. J. Ayer (Ed.) *Logical positivism* (pp. 165–198). Simon and Schuster.

Carnap, R. (1934/1937 [1934]). Logische syntax der sprache. Verlag von Julius springer [*The Logical syntax of language*]. (A. Smeaton, trans.). Harcourt, Brace & Company.

Chalmers, D. (1996). *The conscious mind: In search of a fundamental theory*. Oxford University Press.

Chemero, A. (2009). *Radical embodied cognitive science*. MIT Press.

Chemero, A. (2023). LLMs differ from human cognition because they are not embodied. *Nature Human Behaviour, 7*, 1828–1829.

Chiesa, M. (1994). *Radical behaviorism: The philosophy and the science*. Authors Cooperative.

Churchland, P. M. (1981). Eliminative materialism and the propositional attitudes. *The Journal of Philosophy, 78*(2), 67–90.

Churchland, P. M. (1989). *A neurocomputational perspective: The nature of mind and the structure of science*: MIT Press.

Churchland, P. M. (1995). *The engine of reason, the seat of the soul: A philosophical journey into the brain*. MIT Press.

Churchland, P. M. (1984/2013). Matter and consciousness: A contemporary introduction to the philosophy of mind (3rd ed.). MIT Press.

Churchland, P. S. (1986). *Neurophilosophy: Toward A unified science of the mind-brain*. MIT Press.

Churchland, P. M., & Churchland, P. S. (1990). Could a machine think? *Scientific American, 262*(1), 32–39.

Churchland P. S., Ramachandran, V. S., & Sejnowski, P. (1994). A critique of pure vision. In C. Koch & J. L. Davis (Eds.), *Large-scale neuronal theories of the brain* (pp. 23–60). MIT Press.

Cirera, R. (1993). Carnap's philosophy of mind. *Studies in History and Philosophy of Science Part A, 24*(3), 351–358.

Chomsky, N. (1957). *Syntactic structures*. De Gruyter Mouton.

Chomsky, N. (1959). A review of BF skinner's verbal behavior. *Language, 35*(1), 26–58.

Chomsky, N. (1966). *Cartesian linguistics*. Harper and Row.

Chomsky, N. (1980). *Rules and representations*. Oxford University Press.

Clark, A. (1997). *Being there: Putting brain, body, and world together again*. MIT Press.

Clark, A. (2003). *Natural-born cyborgs: Minds, technologies, and the future of human intelligence*. Oxford University Press.

Clark, A. (2008). *Supersizing the mind: Embodiment, action, and cognitive extension*. Oxford University Press.

Clark, A. (2000/2014). *Mindware. An introduction to the philosophy of cognitive science* (2nd ed.). Oxford University Press.

Clark, A. (2015). *Surfing uncertainty: Prediction, action, and the embodied mind*. Oxford University Press.

Clark, A., & Chalmers, D. (1998). The extended mind. *Analysis, 58*(1), 7–19.

Clark, A., & Toribio, J. (1994). Doing without representing? *Synthese, 101*(3), 401–431.

Clark, A. (1993). *Sensory qualities*. Clarendon.

Clark, A. (2000). *A theory of sentience*. Oxford University Press.

Cobb, M. (2020). *The idea of the brain. A history*. Profile Books.

Collins, G. P. (2002). Claude E. Shannon: Founder of information theory. *Scientific American*, October 14. https://www.scientificamerican.com/article/claude-e-shannon-founder/

Crawford, S. (2013). The myth of logical behaviorism and the origins of the identity theory. In M. Beaney (Ed.), *The Oxford handbook of the history of analytic philosophy* (pp. 621–655). Oxford University Press.

Crawford, S. (2014). On the logical positivists' philosophy of psychology: Laying a legend to rest. In M. C. Galavotti, D. Dieks, W. J. Gonzalez, S. Hartmann, T. Uebel, & M. Weber (Eds.), *New directions in the philosophy of science* (pp. 711–726). Springer.

Danziger, K. (1997). *Naming the mind. How psychology found its language*. Sage Publications.

Darwin, C. (1859). *On the origin of species by means of natural selection, or the preservation of favoured races in the struggle for life*. John Murray.

Darwin, C. (1872). *The expression of the emotions in man and animals*. John Murray.

Darwin, C. (1881). *The formation of vegetable mould, through the action of worms: With observations on their habits*. John Murray.

De Haan, S. (2020). *Enactive psychiatry*. Cambridge University Press.

Democritus. (1951/1999). In *Die Fragmente der Vorsokratiker* (6th ed., Hermann A. Diels & Walther Kranz, Eds.; C. C. W. Taylor, Trans). Weidmann.

The Atomists: Leucippus and Democritus. Fragments, A Text and Translation with Commentary. University of Toronto Press.

Dennett, D. C. (1969). *Content and consciousness*. Routledge.

Dennett, D. C. (1971). Intentional systems. *The Journal of Philosophy*, *68*(4), 87–106.

Dennett, D. C. (1978). *Brainstorms: Philosophical essays on mind and psychology*. MIT Press.

Dennett, D.C. C. (1987). *The intentional stance*. MIT Press.

Dennett, D. C. (1991a). *Consciousness explained*. Little, Brown & Company.

Dennett, D. C (1991b). Real patterns. *Journal of Philosophy*, *88*(1), 27–51.

Dennett, D. C. (1995). *Darwin's dangerous idea. Evolution and the meanings of life*. Simon and Schuster.

Dennett, D. C. (2009). Darwin's "strange inversion of reasoning." *Proceedings of the National Academy of Sciences*, *106*(supplement 1), 10061–10065.

Descartes, René. (1966–1976). *Oeuvres de Descartes* (C. Adam & P. Tannery, Eds.). Vrin/CNRS. Abbreviated as AT in text.

Descartes, René. (1984). *The philosophical writings of Descartes* (vols. I and II; J. Cottingham, R. Stoothoff, & D. Murdoch, Eds., Trans.). Cambridge University Press. Abbreviated as CSM1 and CSM2 in text.

Dewey, J. (1896). The reflex arc concept in psychology. *Psychological Review*, *3*, 357–370.

Dewey, J. (1910). *The influence of Darwin on philosophy and other essays*. Henry Holt and Company.

Dewey, J. (1922). *Human nature and conduct. An introduction to social psychology*. Henry Holt and Company.

Di Paolo, E. A. (2005). Autopoiesis, adaptivity, teleology, agency. *Phenomenology and the Cognitive Sciences*, *4*(4), 429–452.

Di Paolo, E. A., Buhrmann, T., & Barandiaran, X. (2017). *Sensorimotor life: An enactive proposal*. Oxford University Press.

Di Paolo, Ezequiel A., Cuffari, E. C., & De Jaegher, H. (2018). *Linguistic bodies: The continuity between life and language*. MIT Press.

Di Paolo, E. A., & Thompson, E. (2014). The enactive approach. In L. A. Shapiro (Ed.), *The Routledge handbook of embodied cognition* (pp. 68–78). Routledge, Taylor & Francis Group.

Dixon, J. A., Kelty-Stephen, D. G., & Anastas, J. (2014). The embodied dynamics of problem solving: New structure from multiscale interactions. In L. A. Shapiro (Ed.), *The Routledge handbook of embodied cognition* (pp. 160–170). Routledge, Taylor & Francis Group.

Dijksterhuis, E. J. (1950/1961). *De mechanisering van het wereldbeeld.* Meulenhoff. Translated as *The Mechanization of the world picture* (C. Dikshoorn, Trans.). Oxford University Press.

Dijkstra, K., Kaschak, M. P., & Zwaan, R. (2007). Body posture facilitates retrieval of autobiographical memories. *Cognition, 102*(1), 139–149.

Dretske, F. (1986). Misrepresentation. In R. Bogdan (Ed.), *Belief: Form, content, and function* (pp. 17–36). Oxford University Press.

Dretske, Fred. (1988). *Explaining behavior: Reasons in a world of causes.* MIT Press.

Dreyfus, H. L. (1972). *What computers can't do: A critique of artificial reason.* Harper & Row.

Dreyfus, Hubert L. (1992). *What computers still can't do: A critique of artificial reason.* MIT Press.

Evans, G. (1982). *The varieties of reference* (J. McDowell, Ed.). Clarendon Press.

Farrell, B. A. (1950). Experience. *Mind, 59*, 170–198.

Fechner, G. (1860). *Elemente der psychophysik* (2 vol.). Breitkopf und Härtel.

Feyerabend, P. K. (1963). Comment: Mental events and the brain. *The Journal of Philosophy, 60*(11), 295–296.

Finger, S. (1994). *Origins of neuroscience: Explorations into brain function.* Oxford University Press.

Fodor, J. A. (1974). Special sciences. *Synthese, 28*, 97–115.

Fodor, J. A. (1975). *The language of thought.* Crowell.

Fodor, J. A. (1980). Methodological solipsism considered as a research strategy in cognitive psychology. *Behavioral and Brain Sciences, 3*(1), 63–73.

Fodor, J. A. (1983). *Representations: Philosophical essays on the foundations of cognitive science.* MIT Press.

Fodor, J. A. (1990). *A theory of content and other essays*. MIT Press.

Fodor, J. A. (2008). *LOT 2: The language of thought revisited*. Oxford University Press.

Fodor, J. A., & Pylyshyn, Z. W. (1988). Connectionism and cognitive architecture: A critical analysis. *Cognition, 28*(1–2), 3–71.

Fodor, J. A., & Pylyshyn, Z. W. (2015). *Minds without meanings: An essay on the content of concepts*. MIT Press.

Frankish, K. (2016). Illusionism as a theory of consciousness. *Journal of Consciousness Studies, 23*(11–12), 11–39.

Freeman, W. J. (2009). Nonlinear dynamics and intention according to Aquinas. *Mind and Matter, 6*(2), 207–234.

Frege, G. (1892/1948) Über Sinn und Bedeutung. *Zeitschrift für Philosophie und philosophische Kritik, 100*, 25–50. Translated as Sense and reference, by Max Black. *Philosophical Review, 57*(3), 209–230.

Galilei, G. (1623/1957). Il Saggiatore [The Assayer]. (S. Drake, Trans.). In *Discoveries and opinions of Galileo*. Anchor Books.

Gallagher, S. (2017). *Enactivist interventions. Rethinking the mind*. Oxford University Press.

Gallagher, S., & Zahavi, D. ([2012] 2021). *The phenomenological mind* (3d ed). Routledge.

Gardner, M. (1985). *The mind's new science: A history of the cognitive revolution*. Basic Books.

Gellatly, A. (1995). Colourful Whorfian ideas: Linguistic and cultural influences on the perception and cognition of colour, and on the investigation of them. *Mind & Language, 10*(3), 199–225.

Gibson, E. J., & Pick, A. D. (2000). *An ecological approach to perceptual learning and development*. Oxford University Press.

Gibson, J. J. (1963). The useful dimensions of sensitivity. *American Psychologist, 18*(1), 1–15.

Gibson, J. J. (1966). *The senses considered as perceptual systems*. Houghton Mifflin Company.

Gibson, J. J. (1979/1986). *The ecological approach to visual perception*. Psychology Press.

Gibson, J. J., & Gibson, E. J. (1955). Perceptual learning: Differentiation or enrichment? *Psychological Review, 62*(1), 32–41.

Glenberg, A. M., & Kaschak, M. P. (2002). Grounding language in action. *Psychonomic Bulletin & Review, 9*(3), 558–565.

Godfrey-Smith, P. (2020). *Metazoa: Animal minds and the birth of consciousness*. William Collins.

Goethe, J.W. (1808/1962). *Faust. Der Tragödie erster Teil*. Translated as Faust: Part one. Translator: Bayard Taylor. Editor: Stuart Atkins. Collins-MacMillan 1962.

Goff, P. (2019). *Galileo's error: Foundations for a new science of consciousness*. Pantheon books.

Goodman, N. (1968). *Languages of art: An approach to a theory of symbols*. The Bobbs-Merrill Company.

Grice, H. P. (1957). Meaning. *The philosophical review, 66*(3), 377–388.

Griffiths, P. E. (2001). Genetic information: A metaphor in search of a theory. *Philosophy of Science, 68*(3), 394–412.

Haken, H., Kelso, J. A. S., & Bunz, H. (1985). A theoretical model of phase transitions in human hand movements. *Biological Cybernetics, 51*(5), 347–356.

Hanfling, O. (1981). *Logical positivism*. Columbia University Press.

Harrison, P. (1992). Descartes on animals. *The Philosophical Quarterly, 42*(167), 219–227.

Haugeland, J. (2000). *Having thought: Essays in the metaphysics of mind*. Harvard University Press.

Hatfield. G. (2019). Mind and psychology in Descartes. In S. Nadler, T. M. Schmaltz, & D. Antoine-Mahut (Eds.), *The Oxford handbook of Descartes and Cartesianism* (pp. 106–123). Oxford University Press.

Hatfield, G. (2021). Sense data. In E. N. Zalta (Ed.), *The Stanford encyclopedia of philosophy* (Fall 2021 ed.). https://plato.stanford.edu/archives/fall2021/entries/sense-data/

Hauk, O., Johnsrude, I., & Pulvermüller, F. (2004). Somatotopic representation of action words in human motor and premotor cortex. *Neuron, 41*(2), 301–307.

Heidegger, M. (1927/1962). *Sein und zeit* [*Being and time*]. Max Niemeyer Verlag. (J. Macquarrie, & E. Robinson, Trans.). SCM Press.

Hempel, C. G. (1935/1949). Analyse logique de la psychologie. *Revue de Synthese, 10*, 27–42. Translated and reprinted as The logical analysis of psychology. In H. Feigl, & W. Sellars *Readings in philosophical analysis* (pp. 373–384). Century Crofts Inc.

Henrich, J. (2016). *The secret of our success. How culture is driving human evolution, domesticating our species, and making us smarter*. Princeton University Press.

Hobbes, T. (1651/1994). *Leviathan* (E. Curley, Ed.). Hackett.

Hoehl, S., Keupp, S., Schleihauf, H., McGuigan, N., Buttelmann, D., & Whiten, A. (2019). 'Over-imitation': A review and appraisal of a decade of research. *Developmental Review, 51*, 90–108.

Hohwy, J. (2013). *The predictive mind.* Oxford University Press.

Horner, V., & Whiten, A. (2005). Causal knowledge and imitation/emulation switching in chimpanzees (*Pan troglodytes*) and children (*Homo sapiens*). *Animal cognition, 8*, 164–181.

Hume, D. (1748/1999). *An enquiry concerning human understanding* (T. Beauchamp, Ed.). Clarendon Press.

Hurley, S. (1998). *Consciousness in action.* Harvard University Press.

Hutto, D. D., & Myin, E. (2013). *Radicalizing enactivism. Basic minds without content.* MIT Press.

Hutto, D. D., & Myin, E. (2017). *Evolving enactivism: Basic minds meet content.* MIT Press.

Hutto, D. D., & Robertson, I. (2023). Against intellectualism about skill. *Synthese, 201*(4), 1–20.

Hutto, D. D., & Satne, G. (2015). The natural origins of content. *Philosophia, 43*, 521–536.

Huxley, J. (1942). *Evolution: The modern synthesis.* Allen and Unwin.

Huxley, T. H. (1874). On the hypothesis that animals are automata, and its history. *The Fortnightly Review, 16*, 555–580.

Ianì, F. (2024). Reification and cognitive science. *Journal of Theoretical and Philosophical Psychology.* Advance online publication. https://doi.org/10.1037/teo0000296

Ibbotson, P., & Tomasello, M. (2016). Language in a new key. *Scientific American, 315*(5), 70–75.

Jablonka, E., & Lamb, M. J. (2005/2014). *Evolution in four dimensions, revised edition: Genetic, epigenetic, behavioral, and symbolic variation in the history of life.* MIT Press.

Jackson, F. (1982). Epiphenomenal qualia. *The Philosophical Quarterly, 32*(127), 127–136.

Jacob, P. (2023). Intentionality. In E. N. Zalta, & U. Nodelman (Eds.), *The Stanford encyclopedia of philosophy* (Spring 2023 ed.). https://plato.stanford.edu/archives/spr2023/entries/intentionality/

Jensen, R. (2006). Behaviorism, latent learning, and cognitive maps: Needed revisions in introductory psychology textbooks. *The Behavior Analyst, 29*(2), 187–209.

Johnson, M., & Lakoff, G. (2002). Why cognitive linguistics requires embodied realism. *Cognitive Linguistics, 13*(3), 245–263.

Juarrero, A. (1999). *Dynamics in action*. MIT Press.

Juarrero, A. (2023). *Context changes everything: How constraints create coherence*. MIT Press.

Kant, I. (1781/1999). *Kritik der reinen vernunft* [*Critique of pure reason*], edited and translated by Paul Guyer and Allen W. Wood. Cambridge University Press.

Kant, I. (1790/2001). *Kritik der urteilskraft* [*Critique of the power of judgment*], translated by

Paul Guyer and Eric Matthews (Translator). Cambridge University Press.

Kant, I. (1798/2006). *Die anthropologie in pragmatischer Hinsicht [Anthropology from a Pragmatic Point of View]*, edited by Manfred Kuehn and edited and translated by Robert B. Loude. Cambridge University Press.

Kaschak, M. P., Jones, J. L., Carranza, J., & Fox, M. R. (2014). Embodiment and language comprehension. In L. A. Shapiro (Ed.), *The Routledge handbook of embodied cognition* (pp. 118–225). Routledge, Taylor & Francis Group.

Käufer, S., & Chemero, A. (2015/2021). *Phenomenology: An introduction* (2nd ed.). Polity Press.

Keijzer, F. (2015). Moving and sensing without input and output: Early nervous systems and the origins of the animal sensorimotor organization. *Biology & Philosophy, 30*(3), 311–331.

Kelso, J. A. (1995). *Dynamic patterns: The self-organization of brain and behavior*. MIT Press.

Kim, J. (1996/2011). *Philosophy of mind* (3rd ed.). Routledge.

Kirsh, D., & Maglio, P. (1994). On distinguishing epistemic from pragmatic action. *Cognitive Science, 18*(4), 513–549.

Köhler, W. ([1917]1921/1925). *The mentality of apes* (translated from the 2nd German edition by Ella Winter). Harcourt, Brace and World.

Kuhn, T. S. (1962). *The structure of scientific revolutions*. University of Chicago Press.

Lakoff, G., & Johnson, M. (1980). *Metaphors we live by*. University of Chicago Press.

Lakoff, G., & Johnson, M. (1999). *Philosophy in the flesh. The embodied mind and its challenge to Western thought*. Basic Books.

Lakoff, G., & Núñez, R. E. (2000). *Where mathematics comes from: How the embodied mind brings mathematics into being*. Basic Books.

Lala, K. N., Uller, T., Feiner M. N., Feldman, M. W., & Gilbert, S. (2024). *Evolution evolving: The developmental origins of adaptation and biodiversity*. Princeton University Press.

Laland, K. N. (2017). *Darwin's unfinished symphony: How culture made the human mind*. Princeton University Press.

Lee, D. N. (1980). The optic flow field: The foundation of vision. *Philosophical Transactions of the Royal Society B: Biological Sciences, 290*(1038), 169–179.

Lee, D. N., & Reddish, P. E. (1981). Plummeting gannets: A paradigm of ecological optics. *Nature, 293*(5830), 293–294.

Lee, V. L. (1988). *Beyond behaviorism*. Lawrence Erlbaum Associates.

Levine, J. (1983). Materialism and qualia: The explanatory gap. *Pacific Philosophical Quarterly, 64*, (October), 354–361.

Lewis, D. (1966). An argument for the identity theory. *The Journal of Philosophy, 63*(1), 17–25.

Lewis, D. (1970). How to define theoretical terms. *The Journal of Philosophy, 67*(13), 427–446.

Lewis, D. (1972). Psychophysical and theoretical identifications. *Australasian Journal of Philosophy, 50*(3), 249–258.

Lewis, D. (1988/1999). What experience teaches. *Proceedings of the Russellian Society, University of Sydney, 13*, 29–57. Reprinted in Lewis, D. (1999). *Papers in metaphysics and epistemology* (pp. 262–290). Cambridge University Press.

Lewontin, R. C. (1970). The units of selection. *Annual Review of Ecology and Systematics, 1*,1–18.

Lewontin, R. C. (2000). *The triple helix: Gene, organism, and environment*. Harvard University Press.

Lindberg, D.C. (1976). *Theories of vision from Al-Kindi to Kepler*. University of Chicago Press.

Lindberg, D. C. (1992). *The beginnings of Western science: The European scientific tradition in philosophical, religious, and institutional context, prehistory to A.D. 1450*. University of Chicago Press.

Locke, J. (1690/1975). *An essay concerning human understanding* (P. H. Nidditch, Ed.). Oxford University Press.

Lokhorst, G. (2021). Descartes and the pineal gland. In E. N. Zalta (Ed.), *The Stanford encyclopedia of philosophy* (Winter 2021 ed.). https://plato.stanford.edu/archives/win2021/entries/pineal-gland/

Lycan, W. G. (Ed.). (1990). *Mind and cognition: An anthology*. Blackwell Publishing.

Mac Cumhaill, C., & Wiseman, R. (2022). *Metaphysical animals. How four women brought philosophy back to life*. Chatto & Windus.

MacCorquodale, K. (1969). B. F. Skinner's verbal behavior: A retrospective appreciation. *Journal of the Experimental Analysis of Behavior, 12*(5), 831–841.

MacCorquodale, K. (1970). On Chomsky's review of Skinner's verbal behavior. *Journal of the Experimental Analysis of Behavior, 13*(1), 83–99.

Maclaren, K. (2002). Intercorporeality, intersubjectivity and the problem of "letting others be." *Chiasmi International, 4*, 187–210.

Mace, W. M. (1977). James J. Gibson's strategy for perceiving: Ask not what's inside your head, but what your head is inside of. In S. Robert & J. Bransford (Ed.), *Perceiving, acting, and knowing: Toward an ecological psychology* (pp. 43–65). Erlbaum.

Marks, L. E. (2011). A brief history of sensation and reward. In J. A. Gottfried (Ed.) *Neurobiology of sensation and reward* (pp. 15–44). Taylor & Francis.

Marr, D. (1980). *Vision: A computational investigation into the human representation and processing of visual information*. MIT Press.

McAlister, L. L. (ed.). *Psychology from an Empirical Standpoint*, translated by A. C. Rancurello, D. B. Terrell, and L. L. McAlister. New York: Humanities Press, 1973.

Menary, R. (2007). *Cognitive integration, mind and cognition unbounded*. Springer.

Merleau-Ponty, M. (1945/2012). *Phénomenologie de la perception* [*Phenomenology of perception*] (D. Landes, Trans.). Routledge.

Mill, J. S. (1842/1864). Bailey on Berkeley's theory of vision. *Westminster Review*. Reprinted in Mill, J. S. (1864). *Dissertations and discussions: Political, philosophical, and historical* (vol. II, pp. 162–191). Spencer.

Mill, J. S. (1843). *System of logic*. John W. Parker.

Millikan, R. G. (1984). *Language, thought, and other biological categories: New foundations for realism*. MIT Press.

Millikan, R. G. (1993). *White queen psychology and other essays for Alice*. MIT Press.

Minsky, M., & Papert, S. A. (1969). *Perceptrons: An introduction to computational geometry*. MIT Press.

Mook, D. M., Jeffrey, J., & Neuringer, J. (1993). Spontaneously hypertensive rats (SHR) readily learn to vary but not repeat instrumental responses. *Behavioral and Neural Biology, 59*(2), 126–135.

Moyal-Sharrock, D. (2000). Words as deeds: Wittgenstein's "spontaneous utterances" and the dissolution of the explanatory gap. *Philosophical Psychology, 13*(3), 355–372.

Moyal-Sharrock, D. (2004). *Understanding Wittgenstein's* On Certainty. Palgrave.

Moyal-Sharrock, D. (2013). Wittgenstein's razor: The cutting edge of enactivism. *American Philosophical Quarterly, 50*(3), 263–279.

Müller, J. (1833–1840). *Handbuch der physiologie des menschen.* J. Hölscher. Translated as *Elements of Phsyiology* (1837–1843), translated by W. Baly. Taylor & Walton.

Nagel, T. (1974). What is it like to be a bat? *Philosophical Review, 79*, 394–403.

Narvaez, D., Moore, D. S., Witherington, D. C., Vandiver, T. I., & Lickliter, R. (2022). Evolving evolutionary psychology. *American Psychologist, 77*(3), 424–438.

Neisser, U. (1976). *Cognition and reality.* W. H. Freeman.

Neuringer, A. (2002). Operant variability: Evidence, functions, and theory. *Psychonomic Bulletin & Review, 9*(4), 672–705.

Newton, I. (1687/1999). *Philosophiae Naturalis Principia Mathematica* [*Mathematical principles of natural philosophy*]. Translated by I. B. Cohen and A. M. Whitman. Berkeley: University of California Press.

Newton, I. (1730/1952). *Opticks.* Dover.

Nichols, R. (2007). *Thomas Reid's theory of perception.* Clarendon Press.

Nichols, R., & Yaffe, G. (2023). *Thomas Reid.* In E. N. Zalta, & U. Nodelman (Eds.), *The Stanford encyclopedia of philosophy* (Summer 2023 ed.). https://plato.stanford.edu/archives/sum2023/entries/reid/

Noë, A. (2005). Against intellectualism. *Analysis, 65*(4), 278–290.

O'Regan, J. K. (2011). *Why red doesn't sound like a bell: Understanding the feel of consciousness.* Oxford University Press.

O'Regan, J. K., & Noë, A. (2001). A sensorimotor account of vision and visual consciousness. *Behavioral and Brain Sciences, 24*(05), 939–973.

Odling-Smee, J. (2024). *Niche construction: How life contributes to its own evolution.* MIT Press.

Oyama, S. (1985). *The ontogeny of information: Developmental systems and evolution.* Duke University Press.

Page, S., & Neuringer, A. (1985). Variability is an operant. *Journal of Experimental Psychology: Animal Behavior Processes, 11*(3), 429–452.

Patton, L. (2018). Helmholtz's physiological psychology. In S. Lapointe (Ed.), *Philosophy of mind in the nineteenth century. The history of the philosophy of mind* (vol. 5, pp. 96–116). Routledge.

Place, U. T. (1956). Is consciousness a brain process? *British Journal of Psychology, 47*(1), 44–50.

Place, U. T. (1988). Skinner's distinction between rule-governed and contingency-shaped behaviour. *Philosophical Psychology, 1*(2), 225–234.

Piccinini, G. (2015). *Physical computation: A mechanistic account.* Oxford University Press.

Polger, T. W. (2004). *Natural minds.* MIT Press.

Polger, T. W., & Shapiro, L. A. (2016). *The multiple realization book.* Oxford University Press.

Popp, J. A. (2007). *Evolution's first philosopher: John Dewey and the continuity of nature.* State University of New York Press.

Pray, L. (2008). Discovery of DNA structure and function: Watson and Crick. *Nature Education, 1*(1), 100.

Pucetti, R. (1977). The great c-fiber myth: A critical note. *Philosophy of Science, 44*(2), 303–305.

Putnam, H. (1960). Minds and machines. In S. Hook (Ed.), *Dimensions of mind: A symposium* (pp. 138–164). New York University Press.

Putnam, H. (1965). Brains and behavior. In R. J. Butler (Ed.), *Analytical philosophy* (2nd Series, pp. 1–19). Blackwell Publishing.

Putnam H. (1975). Philosophy and our mental life. In H. Putnam (Ed.), *Mind, language, and reality* (pp. 291–303). Cambridge University Press.

Rand, G., & Wapner, S. (1967). Postural status as a factor in memory. *Journal of Verbal Learning and Verbal Behavior, 6*(2), 268–271.

Reed, E. S. (1982). Descartes' corporeal ideas hypothesis and the origin of scientific psychology. *The Review of Metaphysics, 35*(4), 731–754.

Reed, E. S. (1996). *Encountering the world: Toward an ecological psychology.* Oxford University Press.

Reed, E. S. (1997). *From soul to mind: The emergence of psychology, from Erasmus Darwin to William James.* Yale University Press.

Reid, T. (1764/2000). *An inquiry into the human mind on the principles of common sense,* edited by D. R. Brooks. Edinburg University Press.

Reid, T. (1785/2002). *Essays on the intellectual powers of man* (D. R. Brooks & K. Haakonson, Eds.). Edinburg University Press.

Richards, G. (1992). *Mental machinery: The origins and consequences of psychological ideas* (vol. 1, pp. 1600–1850). Athlone Press.

Richards, R. J. (1987). *Darwin and the emergence of evolutionary theories of mind and behavior.* University of Chicago Press.

Rietveld, E. (2008). Situated normativity: The normative aspect of embodied cognition in unreflective action. *Mind, 117*(468), 973–1001.

Rietveld, E., & Kiverstein, J. (2014). A rich landscape of affordances. *Ecological Psychology, 26*(4), 325–352.

Riskin, J. (2016). *The restless clock: A history of the centuries-long argument over what makes living things tick.* University of Chicago Press.

Riskind, J. H. (1983). Nonverbal expressions and the accessibility of life experience memories: A congruence hypothesis. *Social Cognition, 2*(1), 62–86.

Rorty, R. (1965). Mind-body identity, privacy, and categories. *The Review of Metaphysics, 19*(1), 24–54.

Rorty, R. (1979). *Philosophy and the mirror of nature.* Princeton University Press.

Rozemond, M. (1998). *Descartes's dualism.* Harvard University Press.

Rozemond, M. (2011). Real distinction, separability, and corporeal substance in Descartes. *Midwest Studies in Philosophy, 35*(1), 240–258.

Russell, B. (1919). On propositions: What they are and how they mean. *Aristotelian Society Supplementary Volume, 2,* 1–43.

Ryle, G. (1949a). *The concept of mind.* Hutchinson's University Library.

Ryle, G. (1949b). Meaning and necessity. *Philosophy, 24*(88), 69–76.

Ryle, G. (1954). *Dilemmas.* Cambridge University Press.

Ryle, G. (1968/2009). Thinking and reflecting. In Royal Institute of Philosophy (Ed.), *The human agent: Royal institute of philosophy lectures 1966–1967* (pp. 210–226). Macmillan. Reprinted in Ryle, G. (2009). *Collected essays 1929–1968* (pp. 479–493). Routledge.

Schwenkler, J. (2019). *Anscombe's intention: A guide.* Oxford University Press.

Searle, J. R. (1980). Minds, brains, and programs. *Behavioral and Brain Sciences, 3*(3), 417–457.

Segundo-Ortin, M., & Raja, V. (2024). *Ecological psychology.* Cambridge University Press.

Sellars W. (1956). *Empiricism and the philosophy of mind.* In H. Feigl (Ed.), *Minnesota studies in the philosophy of science* (vol. 1, pp. 253–329). University of Minnesota Press.

Shakespeare, W. (1623/2007). *King Lear,* edited by Burton Raffel. Yale University Press.

Shanon, B. (1988). Semantic representation of meaning: A critique. *Psychological Bulletin, 104*(1), 70–83.

Shannon, C. E. (1948). A mathematical theory of communication. *The Bell System Technical Journal, 27*(3), 379–423.

Shannon, C. E. (1950). Programming a computer for playing chess. *The London, Edinburgh, and Dublin Philosophical Magazine and Journal of Science, 41*(314), 256–275.

Shapin, S. (1996). *The scientific revolution.* University of Chicago Press.

Shapiro, L. E., & Spaulding, S. (2024). *The Routledge handbook of embodied cognition* (2nd ed.). Routledge.

Shapiro, L. (2007). *The correspondence between princess Elisabeth of Bohemia and René Descartes.* University of Chicago Press.

Shields, C. (2020). Aristotle's psychology. In E. N. Zalta (Ed.), *The Stanford encyclopedia of philosophy* (Winter 2020 ed.). https://plato.stanford.edu/archives/win2020/entries/aristotle-psychology/.

Simons, D. J., & Chabris, C. F. (1999). Gorillas in our midst: Sustained inattentional blindness for dynamic events. *Perception, 28*(9), 1059–1074.

Skinner, B. F. (1938). The behavior of organisms. Appleton-Century.

Skinner, B. F. (1953). *Science and human behavior.* Macmillan.

Skinner, B. F. (1957). *Verbal behavior.* Appleton-Century-Crofts.

Skinner, B. F. (1963). Behaviorism at fifty. *Science, 140*(3570), 951–958.

Skinner, B. F. (1966). An operant analysis of problem solving. In B. Kleinmuntz (Ed.), *Problem solving: Research, method and theory* (pp. 225–257). John Wiley & Sons.

Skinner, B. F. (1981). Selection by consequences. *Science, 213*(4507), 501–504.

Smart, J. J. (1959). Sensations and brain processes. *The Philosophical Review, 68*(2), 141–156.

Stampe, D. W. 1977. Toward a causal theory of linguistic representation. *Midwest Studies in Philosophy, 2*, 42–63.

Stanley, J., & Williamson, T. (2001). Knowing how. *The Journal of Philosophy, 98*(8), 411–444.

Steiner, P. (2018). Reading and understanding: On some differences between Wittgenstein and 4E cognitive science. *Rivista Italiana di Filosofia del Linguaggio, 12*(2), 124–137.

Stuart-Kennedy, M., & Terrace, H. (2017). In the beginning: A review of Robert C. Berwick and Noam Chomsky's Why Only Us. *Journal of Language Evolution, 2*(2), 114–125.

Tanney, J. (2009). Rethinking Ryle. In G. Ryle (Ed.), *Collected essays 1929–1968* (pp. ix–lvii). Routledge.

Tanney, J. (2022). Gilbert Ryle. In E. N. Zalta (Ed.), *The Stanford encyclopedia of philosophy* (Summer 2022 ed.). https://plato.stanford.edu/archives/sum2022/entries/ryle/

Ter Hark, M. (1990). *Beyond the inner and the out: Wittgenstein's philosophy of psychology*. Springer.

Ter Hark, M. (2009). Coloured vowels: Wittgenstein on synaesthesia and secondary meaning. *Philosophia, 37*, 589–604.

Thelen, E., & Fisher, D. M. (1982). Newborn stepping: An explanation for a "disappearing" reflex. *Developmental Psychology, 18*(5), 760–775.

Thelen, E., Fisher, D. M., & Ridley-Johnson, R. (1984). The relationship between physical growth and a newborn reflex. *Infant Behavior and Development, 7*(4), 479–493.

Thelen, E., & Smith, L. B. (1994). *A dynamic systems approach to the development of cognition and action*. MIT Press.

Thelen, E., & Smith, L. B. (2006). Dynamic systems theories. In W. Damon, & R. M. Lerner (Eds.), *Handbook of child psychology: Vol. 1: Theoretical models of human development* (6th ed., pp. 258–312; Richard M. Lerner, Series Ed.). John Wiley & Sons.

Thelen, E., Schöner, G., Scheier, C., & Smith, L. B. (2001). The dynamics of embodiment: A field theory of infant perseverative reaching. *Behavioral and Brain Sciences, 24*(1), 1–34.

Thompson, E. (2007). *Mind in life: Biology, phenomenology, and the sciences of mind*. Harvard University Press.

Thompson, R. K., Oden, D. L., & Boysen, S. T. (1997). Language-naive chimpanzees (Pan troglodytes) judge relations between relations in a conceptual matching-to-sample task. *Journal of Experimental Psychology: Animal Behavior Processes, 23*(1), 31.

Thorndike E. L. (1911). *Animal intelligence: Experimental studies*. Macmillan.

Thornton, T. (1998). *Wittgenstein on language and thought: The philosophy of content*. Edinburgh University Press.

Titchener, E. B. (1905). *Experimental psychology: A manual of laboratory practice* (vol. II: Quantitative experiments). Macmillan.

Tolman, E. C. (1948). Cognitive maps in animals and man. *Psychological Review, 55*, 189–208.

Tolman, E. C., & Honzik, C. H. (1930a). Degrees of hunger, reward and non-reward, and maze learning in rats. *University of California Publications in Psychology, 4*, 241–256.

Tolman, E. C., & Honzik, C. H. (1930b). *Introduction and removal of reward, and maze performance in rats* (vol. 4, pp. 257–275). University of California Publications in Psychology.

Tolman, E. C., Ritchie, B. F., & Kalish, D. (1946). Studies in spatial learning. I. Orientation and the short-cut. *Journal of Experimental Psychology, 36*(1), 13–24.

Tomasello, M. (2005). *Constructing a language: A usage-based theory of language acquisition*. Harvard University Press.

Tomasello, M., Hare, B., Lehmann, H., & Call, J. (2007). Reliance on head versus eyes in the gaze following of great apes and human infants: The cooperative eye hypothesis. *Journal of Human Evolution, 52*(3), 314–320.

Travis, C. (2013). *Perception: Essays after Frege*. Oxford University Press.

Turing, A. M. (1937). On computable numbers, with an application to the Entscheidungs problem. *Proceedings of the London Mathematical Society, s2-42*(1), 230–265.

Uzgalis, W. (2022). John Locke. In Edward N. Zalta & Uri Nodelman (Eds.), *The Stanford Encyclopedia of Philosophy* (Fall 2022 ed.). https://plato.stanford.edu/archives/fall2022/entries/locke/.

van Dijk, L. (2016). Laying down a path in talking. *Philosophical Psychology, 29*(7), 993–1003.

van Dijk, L. (2021). Temporalizing ontology: A case for pragmatic emergence. *Synthese, 198*, 9021–9034.

van Dijk, L., & Withagen, R. (2016). Temporalizing agency: Moving beyond on-and offline cognition. *Theory & Psychology, 26*(1), 5–26.

Van Gelder, T. (1995). What might cognition be, if not computation? *The Journal of Philosophy, 92*(7), 345–381.

Van Orden, G. C., Pennington, B. F., & Stone, G. O. (2001). What do double dissociations prove? *Cognitive Science, 25*(1), 111–172.

Varela, F. J., Thompson, E., & Rosch, E. (1991). *The embodied mind. cognitive science and human experience*. MIT Press.

Veit, W. 2023. *A philosophy for the science of animal consciousness*. Routledge.

Von Förster, H. (1960). On self-organizing systems and their environments. In M. C. Yovits & S. Cameron (Eds.), *Self-organizing systems* (pp. 31–50). Pergamon Press.

Waddington, C. H. (1957). The strategy of the genes*: A discussion of some aspects of theoretical biology*. Allen and Unwin.

Warren, W. H. (1984). Perceiving affordances: Visual guidance of stair climbing. *Journal of Experimental Psychology: Human Perception and Performance, 10*(5), 683–703.

Weber, A., & Varela, F. J. (2002). Life after Kant: Natural purposes and the autopoietic foundations of biological individuality. *Phenomenology and the Cognitive Sciences, 1*(2), 97–125.

West-Eberhard, M. J. (2003). *Developmental plasticity and evolution*. Oxford University Press.

Wiener, N. (1948/1961). *Cybernetics or control and communication in the animal and the machine* (2nd ed.). MIT Press.

Williams, M. (1983). Wittgenstein on representation, privileged objects, and private languages. *Canadian Journal of Philosophy, 13*(1), 57–78.

Wilson, M. D. (1978). *Descartes*. Routledge.

Winograd, T., & Flores, F. (1986). *Understanding computers and cognition: A new foundation for design*. Ablex Publishing Corporation.

Wiseman, R. (2016). *Routledge philosophy guidebook to Anscombe's intention*. Routledge.

Withagen, R. (2022). *Affective Gibsonian psychology*. Routledge.

Wittgenstein, L. (1922). *Tractatus Logico-Philosophicus*. Many Editions.

Wittgenstein, L. (1953). *Philosophische Untersuchungen/Philosophical investigations* (German-English ed., G. E. Anscombe, Trans.). Blackwell Publishing.

Wittgenstein, L. (1969). *Uber Gewissheit/On certainty* (German-English ed., G. E. Anscombe & D. Paul, Trans.). Blackwell Publishing.

Wittgenstein, L. (1980a), *Remarks on the philosophy of psychology* (vol. 1; G. E. Anscombe & G. H. von Wright, Eds.; G. E. Anscombe, Trans.). Blackwell Publishing.

Wittgenstein, L. (1980b). *Remarks on the philosophy of psychology* (vol. 2, G. H. von Wright & H. Nyman, Eds.; G. Luckhardt & M. Aue, Trans.). Blackwell Publishing.

Wundt, W. (1874/1904). *Grundzüge der Phyiologischen Psychologie* [*Principles of physiological psychology*], edited by Wilhelm Engelmann and translated by E. B. Titchener. Swan Sonneschein.

Zahnoun, F. (2020). Explaining the reified notion of representation from a linguistic perspective. *Phenomenology and the Cognitive Sciences, 19*, 79–96.

Zahnoun, F. (2023). *The embodiment of meaning: Why matter matters for cognition and experience*. Routledge.

Zahnoun, F. (2025). Indormation in the brain. *Theoria*. Online Version of Record before inclusion in an issue. https://doi.org/10.1111/theo.70021

Zilio, D. (2016). Who, what, and when: Skinner's critiques of neuroscience and his main targets. *The Behavior Analyst, 39*, 197–218.

Index